Springer Theses

Recognizing Outstanding Ph.D. Research

For further volumes:
http://www.springer.com/series/8790

Aims and Scope

The series "Springer Theses" brings together a selection of the very best Ph.D. theses from around the world and across the physical sciences. Nominated and endorsed by two recognized specialists, each published volume has been selected for its scientific excellence and the high impact of its contents for the pertinent field of research. For greater accessibility to non-specialists, the published versions include an extended introduction, as well as a foreword by the student's supervisor explaining the special relevance of the work for the field. As a whole, the series will provide a valuable resource both for newcomers to the research fields described, and for other scientists seeking detailed background information on special questions. Finally, it provides an accredited documentation of the valuable contributions made by today's younger generation of scientists.

Theses are accepted into the series by invited nomination only and must fulfill all of the following criteria

- They must be written in good English.
- The topic should fall within the confines of Chemistry, Physics, Earth Sciences, Engineering and related interdisciplinary fields such as Materials, Nanoscience, Chemical Engineering, Complex Systems and Biophysics.
- The work reported in the thesis must represent a significant scientific advance.
- If the thesis includes previously published material, permission to reproduce this must be gained from the respective copyright holder.
- They must have been examined and passed during the 12 months prior to nomination.
- Each thesis should include a foreword by the supervisor outlining the significance of its content.
- The theses should have a clearly defined structure including an introduction accessible to scientists not expert in that particular field.

Cornelius Krull

Electronic Structure of Metal Phthalocyanines on Ag(100)

Doctoral Thesis accepted by
Autonomous University of Barcelona, Spain

Author
Dr. Cornelius Krull
Catalan Institute of Nanoscience and
  Nanotechnology (ICN2)
Autonomous University of Barcelona
Bellaterra (Barcelona)
Spain

Supervisor
Dr. Aitor Mugarza
Catalan Institute of Nanoscience and
  Nanotechnology (ICN2)
Autonomous University of Barcelona
Bellaterra (Barcelona)
Spain

ISSN 2190-5053   ISSN 2190-5061 (electronic)
ISBN 978-3-319-02659-6   ISBN 978-3-319-02660-2 (eBook)
DOI 10.1007/978-3-319-02660-2
Springer Cham Heidelberg New York Dordrecht London

Library of Congress Control Number: 2013951128

© Springer International Publishing Switzerland 2014
This work is subject to copyright. All rights are reserved by the Publisher, whether the whole or part of the material is concerned, specifically the rights of translation, reprinting, reuse of illustrations, recitation, broadcasting, reproduction on microfilms or in any other physical way, and transmission or information storage and retrieval, electronic adaptation, computer software, or by similar or dissimilar methodology now known or hereafter developed. Exempted from this legal reservation are brief excerpts in connection with reviews or scholarly analysis or material supplied specifically for the purpose of being entered and executed on a computer system, for exclusive use by the purchaser of the work. Duplication of this publication or parts thereof is permitted only under the provisions of the Copyright Law of the Publisher's location, in its current version, and permission for use must always be obtained from Springer. Permissions for use may be obtained through RightsLink at the Copyright Clearance Center. Violations are liable to prosecution under the respective Copyright Law.
The use of general descriptive names, registered names, trademarks, service marks, etc. in this publication does not imply, even in the absence of a specific statement, that such names are exempt from the relevant protective laws and regulations and therefore free for general use.
While the advice and information in this book are believed to be true and accurate at the date of publication, neither the authors nor the editors nor the publisher can accept any legal responsibility for any errors or omissions that may be made. The publisher makes no warranty, express or implied, with respect to the material contained herein.

Printed on acid-free paper

Springer is part of Springer Science+Business Media (www.springer.com)

Supervisor's Foreword

Molecular materials promise many advantages for applications in technological devices. In electronic, spintronic or optic systems they can provide low cost alternatives to existing approaches and new functionalities that are not accessible with the current inorganic semiconductor technology. Fundamental research in this field has been specially focused on understanding the physics at the molecule-metal interface, motivated by the dramatic effect of the interfacial structure on the transport properties of the film. The reason for that is that the strong interaction needed to have good contact with metallic electrodes perturbs the electronic structure of the interfacial molecular layer. The situation becomes far more critical if we consider single-molecule junctions as possible future devices. Here, the whole molecule-metal interface becomes the active component of the device and can exhibit radically different properties from that of the isolated molecule. Hence, molecular devices with predictable properties will only be realized after gaining an atomic scale understanding of the correlation between structural, electronic, and magnetic properties.

Scanning tunneling microscopy, with the ability to simultaneously perform imaging and local spectroscopy at the atomic scale, is a perfect tool for this type of investigation. A series of fundamental quantum properties can be explored and manipulated by this technique. Elastic spectroscopy provides information on the energy and spatial distribution of the electronic states. In other words, it can be used to map electron wave functions (such as molecular orbitals) at selected energies. The inelastic contribution, on the other hand, can be used to explore excitations of different origin. Finally, the tip can also be exploited as a manipulation tool to realize artificial configurations and structures and to induce chemical reactions at a single-molecule level. Combining all these capabilities, a comprehensive set of information can be obtained on the molecular properties and correlated to its precise chemical state and bonding configuration at the metallic interface.

The work presented in this thesis provides a detailed picture of how the molecule-metal interaction does not only distort molecular orbitals but can also induce a pool of new physical phenomena that determine the transport properties of molecular thin films and junctions. The spectroscopic measurements reveal that chirality can be induced in molecular orbitals by the underlying metallic surface without distorting the molecular conformation. Depending whether the interaction occurs via the metal ion or the organic ligand, the electronic structure is modified in a dramatically different way, and magnetic moments are either being quenched or induced. Vibrational and magnetic excitations can coexist in the same molecule and couple to many-body interactions, fully determining the low-energy transport properties of the junction. These new charge and spin configuration of individual molecules are further modified by using the STM tip to form intermolecular bonds or electron dope by using alkali atoms.

The presented results highlight the need of considering the molecule-metal interface as a functional, active part of molecular junctions.

Barcelona, September 2013   Dr. Aitor Mugarza

Abstract

Traditional semiconductor technology will reach a size limit within the next few years. A possible solution is the use of organic molecules in technological applications, which offer a number of key advantages: their intrinsic small size ($\sim$nm) and ability to self assemble into functional structures. Moreover, the highly developed methods of molecular synthesis allow to tailor their electronic and magnetic properties. However, the implementation of organic devices depends crucially on the understanding of the interaction between molecules and metal electrodes as well as molecule-molecule interactions.

Transition-metal phthalocyanines (MePc) are well-known metal organic complexes, which have bulk semiconducting properties. Their versatile chemistry combined with a relatively simple and robust structure, makes them the ideal system to study the interaction of metal-organic complexes with metal surfaces. This thesis deals with the structural electronic and magnetic properties of MePc molecules adsorbed on a metallic substrate. By using Scanning Tunneling Microscopy (STM) and Scanning Tunneling Spectroscopy (STS), we studied the molecule-substrate and molecule-molecule interaction of four MePc (Me = Fe, Co, Ni, Cu) starting from single molecules up to multilayer films on a Ag(100) surface.

We found that single MePc adsorb in a planar fashion with their molecular axis rotated by $\pm 30°$ with respect to the surface [011] direction. This rotation is caused by the bond optimization between the MePc's aza-N and underlying Ag atoms. Pristine MePc are achiral; however the misalignment between surface and molecular symmetry axis induces the formation of chiral orbitals in the structurally undistorted molecules. For NiPc and CuPc a bias dependent electronic chirality is observed. This effect shows that chirality can be manifested exclusively at the electronic level due to asymmetric charge transfer between molecules and substrate.

At submonolayer coverages, we observed that the molecules self-assemble into clusters with a 5×5 R37° superstructure, and well-defined organizational chirality. The single molecule chirality is univocally transferred to the supramolecular organization of the clusters by means of attractive vdW interactions. The monolayer consists of large homochiral domains that are delimited by the surface terraces. The formation of homochiral domains starting from a racemic mixture of clusters is due to spontaneous symmetry breaking induced by Ostwald ripening and the thermally induced rotation of the molecular axis, which leads to a switching of the chirality.

By using STS, we investigated the electronic structure of four MePc adsorbed on Ag(100). Our systematic approach allowed us to study the charge transfer and hybridization mechanisms of MePc as a function of increasing occupancy of the 3d metal states. We found that all four MePc receive approximately one electron from the substrate. However, depending on the central metal ion, charge transfer from the substrate has different consequences. In FePc and CoPc, it mixes metal and ligand molecular orbitals, and induces a charge reorganization within the complete molecule. In contrast, in NiPc and CuPc one electron is transferred to a ligand π-orbitals, leaving the d states unperturbed.

dI/dV spectra of CuPc and NiPc recorded at different positions within the molecule provide evidence of an intense zero bias Kondo resonance (T_K = 29 K) corresponding to the orbital extension of the ligand $2e_g$ orbital (gas phase LUMO). This is direct evidence of an additional magnetic moment caused by the transfer of one electron to the π states. We further find that this Kondo resonance couples to internal vibrations of the molecule in both NiPc and CuPc. For CuPc the coexistence of spins at the ion's d and ligand π states leads to a triplet ground state, because of the intramolecular exchange coupling between d and π electrons. Inelastic electron tunneling induces excitations of the higher energy singlet state. For FePc and CoPc, on the other hand, neither Kondo resonances nor inelastic excitations are observed. DFT calculations point to a mixed-valence state for both molecules, caused by the different spatial distribution of the frontier orbitals compared to CuPc and NiPc. For both FePc and CoPc the interaction with the substrate tends to reduce the magnetic moment.

We investigated the influence of intermolecular forces on the electronic and magnetic structure of MePc as a function of their coordination number, by manipulating individual molecules to form artificial clusters. CuPc shows a complex evolution of the electronic structure characterized by an upshift of the $2e_g$ ligand orbital. In contrast, for CoPc clusters no changes occur. The differences may arise from the fact that CuPc interacts with the substrate through the ligand, whereas CoPc interacts through the ion. This makes both the induced charge and adsorption configuration of CuPc more sensitive to intramolecular interactions.

Multilayers films of CuPc grow layer-by-layer up to the fifth. Between the first and second layer there is a 45° rotation of the molecules, which maintain the central ion aligned. Higher layers show a lateral shift towards a ion-N alignment, and a gradually increasing tilt. The electronic structure exhibits a gradual decoupling from the substrate, and evolution towards semiconducting behavior, manifested in the opening of a gap around E_F the coupling to molecular vibrations, and voltage-dependent tunneling barriers.

We studied the possibilities to manipulate the spin and change state of MePc through atom-by-atom doping with alkali electron donors. For CuPc we found that both ligand and metal states can be selectively doped by changing the alkali-molecule bonding configuration. This permits tailoring of the spin (S = 0, 1/2, 1) and charge state (Q = 1, 2), with a single alkali dopant. Controlled manipulation of individual Li atoms further allowed us to explore the sequential charging behavior with up to five additional Li atoms and the effect of short and long range Coulomb repulsion on molecular orbitals. The doping of CuPc monolayer films shows that the electrostatic interaction between Li atoms of neighboring Li@CuPc complexes leads to a crossover of stable configurations depending on the Li dosage, favoring charge transfer to the ion states at high Li coverage.

Finally, we investigated the effect of magnetic dopants such as Fe atoms on NiPc and CuPc. No direct evidence of coupling between the magnetic moments of Fe and CuPc was observed. However, the Kondo temperature of the Fe atoms varies due to changes in their local environment modulated by the presence of MePc.

Overall, these results provide a comprehensive view of the interaction of MePc with metal substrates as well as the magnetic and transport properties of single molecules, clusters, monolayer, and multilayers.

Acknowledgments

First and foremost I would like to express my gratitude to Aitor Mugarza, who has the spirit of a true scientist. This thesis, with its ups and downs would not have been possible without him, his encouragement, dedication, and insight. Eskerrik asko denagatik. I am also very grateful to Pietro Gambardella, who gave me the opportunity to learn and research at the ICN, and always could find some time for me despite his many obligations, and who created an excellent atmosphere in the group thanks to his vision and optimism.

I would also like to thank Richard Korytár, Nicolas Lorente, Roberto Robles, and Pablo Ordejon for performing the DFT calculations, which form an important part of this work, Alexander Krull for providing the drift correction code used in STM measurements, and Nicolas Lorente again for taking the time to discuss with me the odds and ends of the Kondo effect.

I feel fortunate to have been in a research group with such a positive attitude and feeling of companionship. I really felt at home and I would like to say thank you—in no particular order—to all of you: To Gustavo Ceballos for the many scientific and not so scientific discussions we had; to Marc Ollé for the many projects we have talked about (lets hope one of them works), the wee little conflicts we fought, and of course for organizing the calçotadas; to Alberto Lodi-Rizzini for sharing a couple of beam times with me, his frosty cookies, bueno all of his cookies, and making sure we would not starve; to Jerald Kravich for also making sure we would not starve and the many strange and interesting conversations we had (although did not often agree with him); to Sebastian Stepanow; to Timofey Balashov for tightening many a flange together; to Corneliu Nistor among other things for sharing my name; to Mihai Miron for being Mihai; to Kevin Garello for climbing many a meter with me and for shared thoughts and beers (or the other way round); to Jorge "Mr. Wolf" Lobo for bringing more laughter in this world; to Gina Peschel, to Sonia Matencio, to Santos Alvarado, to Sylvie Godey, to Can Onur Avci, and to Raoul Piquerel.

Lastly, but by no means least the people who have made my life in Barcelona so very pleasant, Joan, Fran, Helena, Carlos, Diana, Pablo, Severine, Xabi, James, Kevin, y especialmente Marta por asegurarme y aguantarme :)

Publications

1. C. Krull, A. Mugarza, and P. Gambardella. "**Controlling the Kondo interaction of Fe adatoms on Ag(100) through organic molecules**". in preparation

2. C. Krull, A. Mugarza, and P. Gambardella. "**Evolution of the electronic structure for CuPc on Ag(100) as a function of coverage: from clusters to multilayers**". in preparation

3. C. Krull, R. Robles, A. Mugarza, and P. Gambardella. "**Site- and orbital-dependent charge donation and spin manipulation in electron-doped metal phthalocyanines**". *Nature Materials*, 12 337–343, 2012. doi:10.1038/nmat3547.

4. A. Lodi Rizzini, C. Krull, T. Balashov, A. Mugarza, C. Nistor, F. Yakhou, V. Sessi, S. Klyatskaya, M. Ruben, S. Stepanow, and P. Gambardella. "**Exchange biasing single molecule magnets: Coupling of TbPc$_2$ to antiferromagnetic layers**". *Nano Letters* 12 5703, 2012. doi:10.1021/nl302918d.

5. A. Mugarza, R. Robles, C. Krull, R. Korytar, N. Lorente, S. Stepanow, and P. Gambardella. "**Electronic and magnetic properties of molecule-metal interfaces: transition metal phthalocyanines adsorbed on Ag(100)**". *Physical Review B*, 85(15):155437, April 2012. doi:10.1103/PhysRevB.85.155437.

6. A. Lodi Rizzini, C. Krull, T. Balashov, J. J. Kavich, A. Mugarza, P.S. Miedema, P. K. Thakur, V. Sessi, S. Klyatskaya, M. Ruben, S. Stepanow, and P. Gambardella. "**Coupling single molecule magnets to ferromagnetic substrates**". *Physical Review Letters*, 107(17):177205, October 2011. doi:10.1103/PhysRevLett.107.177205.

7. A. Mugarza, C. Krull, R. Robles, S. Stepanow, G. Ceballos, and P. Gambardella. "**Spin coupling and relaxation inside molecule–metal contacts**". *Nat Commun*, 2:490, October 2011. doi:10.1038/ncomms1497.

8. C. Carbone, S. Gardonio, P. Moras, S. Lounis, M. Heide, G. Bihlmayer, N. Atodiresei, P. H. Dederichs, S. Blügel, S. Vlaic, A. Lehnert, S. Ouazi, S. Rusponi, H. Brune, J. Honolka, A. Enders, K. Kern, S. Stepanow, C. Krull, T. Balashov, A. Mugarza, and P. Gambardella. "**Self-Assembled Nanometer-Scale magnetic networks on surfaces: Fundamental interactions and functional properties**". *Advanced Functional Materials*, 21(7):1212–1228, April 2011. doi:10.1002/adfm.201001325.

9. A. Mugarza, N. Lorente, P. Ordejón, C. Krull, S. Stepanow, M. Bocquet, J. Fraxedas, G. Ceballos, and P. Gambardella. "**Orbital specific chirality and homochiral Self-Assembly of achiral molecules induced by charge transfer and spontaneous symmetry breaking**". *Physical Review Letters*, 105(11):115702, September 2010. doi:10.1103/PhysRevLett.105.115702.

10. C. Carbone, M. Veronese, P. Moras, S. Gardonio, C. Grazioli, P. H. Zhou, O. Rader, A. Varykhalov, C. Krull, T. Balashov, A. Mugarza, P. Gambardella, S. Lebègue, O. Eriksson, M. I. Katsnelson, and A. I. Lichtenstein. "**Correlated electrons step by step: Itinerant-to-Localized transition of Fe impurities in Free-Electron metal hosts**". *Physical Review Letters*, 104(11):117601, March 2010. doi:10.1103/PhysRevLett.104.117601.

11. C. Krull, S. Valencia, J. I. Pascual, and W. Theis. "**Formation of extended straight molecular chains by pairing of thymine molecules on the Ag–Si(111) surface**". *Applied Physics A*, 95(1):297–301, December 2008. doi:10.1007/s00339-008-5040-y.

Contents

1 Introduction: Molecular Electronics 1
   References 4

2 Experimental Techniques 9
   2.1 Scanning Tunneling Microscopy 9
      2.1.1 The Operating Principle 10
      2.1.2 Theoretical Descriptions of the Tunneling Process 12
   2.2 Scanning Tunneling Spectroscopy 16
      2.2.1 Elastic Tunneling Spectroscopy 16
      2.2.2 Inelastic Tunneling Spectroscopy 16
      2.2.3 Lock-in Technique 18
      2.2.4 Background Subtraction 19
      2.2.5 Differential Conductance Maps 21
   2.3 Manipulation Techniques 21
   2.4 Experimental Setup 22
   2.5 Methods 26
   References 27

3 Introduction to the Kondo Effect 31
   3.1 The Kondo Problem 31
   3.2 The Anderson Model 33
      3.2.1 The Anderson Hamiltonian 34
   3.3 Virtual Spin Flips and the Schrieffer: Wolff Transformation 36
   3.4 Formation of the Kondo Singlet 38
      3.4.1 The Spin 1/2 Kondo Effect 38
      3.4.2 The Underscreened Kondo Effect 42
      3.4.3 The Inelastic Kondo Effect 44
   3.5 The Kondo Effect in STM Measurements 44
   References 48

4 Adsorption of Metal Phthalocyanines on Ag(100) 51
   4.1 Chemical Structure of Metal Phthalocyanines 52

	4.2	Adsorption of Single Molecules		52
		4.2.1 Adsorption Configuration		52
		4.2.2 Orbital Specific Electronic Chirality		55
	4.3	Monolayer Growth		57
		4.3.1 Evolution of Chirality in CuPc and CoPc Structures		57
		4.3.2 Supramolecular Structure		61
		4.3.3 Origin of the Transfer of Chirality		63
	4.4	Multilayer Growth		66
	4.5	Summary		68
	References			69
5	**Electronic and Magnetic Properties of MePc on Ag(100)**			73
	5.1	Pristine MePc: Gas Phase Electronic Structure		74
	5.2	Single Molecules: Electronic Structure		76
		5.2.1 Spectroscopy of Molecular Orbitals		77
		5.2.2 DFT: Electronic Structure		80
	5.3	Single Molecules: Magnetic Structure		81
		5.3.1 Kondo Interaction		81
		5.3.2 DFT: Magnetic Structure		89
	5.4	From Clusters to Monolayer		93
		5.4.1 Small Clusters of CuPc		93
		5.4.2 Monolayers of CuPc and CoPc		97
	5.5	CuPc Multilayer		99
		5.5.1 Molecules on Higher Layers: Spectroscopy		101
	5.6	Summary		106
	References			109
6	**Doping of MePc: Alkali and Fe Atoms**			115
	6.1	Electron Doping of MePc		115
		6.1.1 Single MePc Doped with Lithium		116
		6.1.2 Li Doping of a Monolayer of CuPc		129
	6.2	Doping with Fe Atoms		131
	6.3	Summary		138
	References			139
7	**Conclusions and Outlook**			141
	References			144
Appendix CuPc on Au(111)				145

Symbols and Abbreviations

E_F	Fermi level
DFT	Density Functional Theory
GGA	Generalized Gradient Approximation
HOMO	Highest Occupied Molecular Orbital
LDA	Local Density Approximation
LDOS	Local Density of States
LEED	Low Energy Electron Diffraction
LUMO	Lowest Unoccupied Molecular Orbital
MePc	Metal Phthalocyanines
MO	Molecular Orbital
N_a	Aza-Nitrogens
N_p	Pyrole-Nitrogen
NDR	Neagtive Differential Resistance
NRG	Numerical Renormalization Group
Pc	Phthalocyanine
PDOS	Projected Density of States
STM	Scanning Tunneling Microscopy
STS	Scanning Tunneling Spectroscopy
TM	Transitional Metal
UHV	Ultra High Vacuum($p < 10^{-9}$ mBar)
vdW	van der Waals

Chapter 1
Introduction: Molecular Electronics

"**Small is Beautiful**", is a statement coming originally from the world of economics [1]. It nevertheless holds true when it comes to technology. Small is beautiful, because small is fast and small is cheap. In the 1950s the miniaturization of technological appliances, especially electronics, started at a fast pace, creating big markets. Miniaturized sensors, micromechanical devices, and integrated circuits were developed quickly. In microelectronics the number of functional units, i.e., transistors roughly doubled each year, a trend known as Moore's law [2]. The need to create ever smaller functional units in devices became a driving force for the new semiconductor industry. Also in science the focus shifted towards smaller systems. Feynman suggested in 1959, in his groundbreaking talk "There is plenty of room at the bottom" [3], the idea of "manipulating and controlling things in a smaller scale"; He suggested that if we were to achieve control over matter at the atomic or molecular level, development in science and technology would be revolutionarized. Today everyday devices are based on industrial scale processes that pattern matter on a nanometer scale. Traditionally, and very successfully, this creation of "small functional units" is based on technologies that make big structures smaller, a method called "top–down" [4]. Most of these techniques are transformations of old principles, such a lithography, writing or stamping applied to the 100-nm scale. This type of precise nanostructuring is achieved by electron-beam writing [5], and advanced lithography techniques using extreme ultra-violet or even hard x-ray radiation [6]. However, pushing these method to even smaller sizes, involves overcoming wavelength limits, and considerable monetray investments [7–9].

An alternative method, which could be a solution to these problems, consists of assembling functional units out of smaller building blocks in a "bottom–up" fashion [4]. These strategies for the fabrication of nanostructures are based on growth phenomena. Atoms or molecules (or both) are deposited on the substrate and ordered nanoscale structures are formed based on different atomistic processes. The resulting structures will be determined by the interplay of surface diffusion, influx of material and the chemical interaction between elements. Controlling these parameters through temperature, flux, composition and molecular synthesis a wide range of different growth modes and structures can be achieved [8].

One of the most promising classes of materials that can be used in such nanofabrication methods are based on organic molecules. In living organisms molecules are smallest functional units performing an incredible number of different tasks. In technological applications the use of molecules may offer a cheaper alternative over the structuring of inorganic matter. Thicker layers of molecular materials (~100 nm) exploit their semiconducting properties, and the intermolecular interactions within these organic films, such as van der Waals forces, dipolar interactions, donor–acceptor recognition, π-stacking, and hydrogen bonds have been studied widely [10]. Their understanding has led to the creation of a range of different applications: organic-light emitting diodes (OLEDs) [11, 12], already commercialized as AMOLED displays, photovoltaic cells for solar energy generation [13–15], gas sensors [16], and thin-film transistors [17].

But molecules have even more to offer, especially when we focus on on their use as small functional units. Their key advantages include [18]:

Structural perfection. Below a certain temperature threshold, usually at room temperature or higher, molecules are structurally perfect and identical, because chemistry stabilizes their composition.

Small size. Molecules already are in the size scale between 1 and 100 nm, making them good building blocks for the construction of functional nanostructures with accompanying advantages in cost, efficiency, and power dissipation.

Self-assembly and recognition. In nature, molecular interactions lead to the self-assembly of complex and functional superstructures e.g., the DNA molecule. This can be used to create nanoscale structures by self-assembly [19–22]. Molecular recognition leads to changes in electronic behavior, providing the basis for both switching and sensing applications [23].

Dynamical stereochemistry. Many molecules have multiple distinct stable geometric structures or isomers. Such geometric isomers can have distinct optical and electronic properties, which in turn can be used to change the transport properties at the single molecular level, creating a molecular switch, e.g., azobenzene, which can be switched by light [24] and current [25].

Synthetic tailorability. The chemistry of molecular synthesis is highly developed. By varying the composition and geometry, it is possible to tailor make structural, bonding, transport, and optical properties of molecules [26].

Despite these advantages, the experimental and theoretical challenges to create devices based on functional molecules are enormous. The possible use of molecules as single functional units, or as an ordered self assembled layer, depends on the understanding and control of two key aspects: The interaction between the molecules themselves and the interaction with metallic electrodes or supporting substrate. Interesting and useful molecular properties can be destroyed, changed or newly created by both of these interactions. It is thus of paramount interest to study single molecules at a large variety of interfaces as well as their interaction in clusters and layer structures.

1 Introduction: Molecular Electronics

Already in 1974 Aviram and Ratner proposed the fabrication of an electronic rectifier based on a single molecule with a donor, spacer and acceptor structure [27]. Since then a lot of effort has been put into the investigation of molecules as functional electronic units. A large group of experiments undertaken in this sector are based on break junction experiments, which allows the conduction properties of molecules to be accessed. Here the molecules are placed between two electrodes created by the mechanical fracturing of a wire or by electromigration [28]. These type of experiments showcase one of the inherent difficulties when contacting molecules: small changes in the uncontrolled geometry of the electrodes can lead to large changes in the conduction measurements [29]. In spite of these difficulties, the proof of princible realization of single molecules acting as functional units was achieved [30]. Many important fundamental conduction phenomanea for these type of systems were observed, described and understood, such as Coulomb blockade [31], vibronic coupling [32], negative differential resistance [33], and the Kondo effect [33, 34]. The latter is related to the coupling of the spin of conduction electrons to that of a magnetic impurity. Below a critical temperature, the so-called Kondo temperature, the conduction characteristics dramatically change due to a screening of the magnetic moment of the impurity (see Chap. 3).

The inclusion of the spin, in addition to the charge, to store and transport information in in electonic devices was a new and powerful concept, first exploited after the discovery of the giant magneto-resistance effect in the late 1980s. It showed that in anorganic heterogeneous magnetic layer structures the electrical resistance was greatly influenced by the magnetic orientations within the device [35, 36]. This marked the beginning of the field of Spintronics, which addresses the problem of injecting, manipulating and detecting spins in the solid state materials [37]. Organic molecules have have recently emerged as a new and promising platform for spintronics, a field known as Molecular Spintronics [38]. An auspicious pathway to create molecules for this line of research is ligand chemistry. Metal-organic complexes can be constructed to have specific magnetic properties, and their hybrid nature can be used to induce room temperature magnetism by combining local and itinerant spins [39]. The spin states of a molecule can be modified by applying a gate voltage [40] or even by mechanically deforming the molecule inside a break junction [41].

This thesis focuses on metal phthalocyanines (MePc), a well known type of coordination complexes formed by a central metal ion and an organic macrocyclic ligand (Pc). The discovery of these molecules dates back to the turn of the twentieth century, when the purely organic Pc was first observed as a strongly colored reaction byproduct. The introduction of metal ion in the center of the Pc-molecule happened by accident in 1928. In a Scottish dye factory, a reaction vessel cracked and exposed the Pc-reagents to the outer steel piping, leading to the creation of a green–blue material. This material was CuPc, a Pc ligand with an Cu ion in the center. The possible use of CuPc, and MePc in general, as dye pigments became immediately apparent, and industrial production started in 1935 [42]. Academic interest also spurred and by 1936 the structure, synthesis, and capability of phthalocyanines molecules to complex with a wide range of metal ions were described.

More than 70 different metal atoms have been found to coordinate with phthalocyanines, creating a vast range of different electronic and magnetic properties. It is exactly this variety, paired with a simple and robust structure that makes them interesting for technological applications. Semiconducting bulk MePc crystals have already been used in thin films for OLED [43, 44] and solar cell devices [45, 46]. For the same reasons, in fundamental research MePcs have become a model system to study the interaction of metal-organic complexes with metal surfaces. Their flat adsorption geometry facilitates the bonding of both the central TM ion and organic ligands to the substrate, whereas their capability to coordinate many different metal atoms allows for a systematic investigation metal ion-dependent investigation of the magnetic properties.

Here we present a study of the MePc model system on a single crystal Ag(100) surface. Starting at the level of individual molecules, we investigate the impact of molecule–substrate and molecule–molecule interactions on the electronic and magnetic structure, and the self-assembly of MePc. The detailed characterization of larger structures such as clusters, monolayers, and multilayers allows us to shed light on the complex interplay between these interactions. The use of four different MePc (Me = Fe, Co, Ni, Cu) allows us determine the role of the central ion in these processes. In a last step we explore the possibility of manipulating of the spin and charge states by doping the molecules with alkali metal atoms and Fe atoms.

Scanning tunneling microscopy (STM) is an ideal tool to investigate physical and chemical processes at interfaces [47]. STM provides information about the structural, electronic, and to some extent, magnetic properties on a sub-molecular scale. Therefore, in contrast to break junction experiments the molecule-electrode bonding configuration can be determined. Molecular adsorption on polycrystalline Silver was studied as early as 1987 [48–50]. Additionally, the ability of STM to manipulate the investigated system on the nanoscale allows the study of intermolecular interactions through the fabrication of different molecular structures [51].

References

1. E.F. Schumacher, *Small Is Beautiful: Economics as if People Mattered*, 2nd edn. (Harper Perennial, New York, 1989), ISBN 0060916303
2. G. Moore et al., Cramming more components onto integrated circuits. Proc. IEEE **86**(1), 82–85 (1998)
3. R.P. Feynman, There's plenty of room at the bottom (1959), http://calteches.library.caltech.edu/47/2/1960Bottom.pdf
4. B. Cui, *Recent Advances in Nanofabrication Techniques and Applications*, (InTech, Rijeka, 2011), ISBN 978-953-307-602-7. doi:10.5772/859
5. A.E. Grigorescu, C.W. Hagen, Resists for sub-20-nm electron beam lithography with a focus on HSQ: state of the art. Nanotechnol. **20**(29), 292001 (2009). doi:10.1088/0957-4484/20/29/292001
6. T. Ito, S. Okazaki, Pushing the limits of lithography. Nature. **406**(6799), 1027–1031 (2000). doi:10.1038/35023233

7. C. Vieu, F. Carcenac, A. Pépin, Y. Chen, M. Mejias, A. Lebib, L. Manin-Ferlazzo, L. Couraud, H. Launois, Electron beam lithography: resolution limits and applications, Appl. Surf. Sci. **164**(1–4), 111–117 (2000). doi:10.1016/S0169-4332(00)00352--4
8. J.V. Barth, G. Costantini, K. Kern, Engineering atomic and molecular nanostructures at surfaces. Nature **437**(7059), 671–679 (2005). doi:10.1038/nature04166
9. C.A. Mack, Line-edge roughness and the ultimate limits of lithography. Proc. SPIE Adv. Resist Mater. Process. Technol. XXVII **7639**, 763931 (2010). doi:10.1117/12.848236
10. S.R. Forrest, Ultrathin organic films grown by organic molecular beam deposition and related techniques. Chem. Rev. **97**(6), 1793–1896 (1997). doi:10.1021/cr941014o
11. J.H. Burroughes, D.D.C. Bradley, A.R. Brown, R.N. Marks, K. Mackay, R.H. Friend, P.L. Burns, A.B. Holmes, Light-emitting diodes based on conjugated polymers. Nature **347**, 539–541 (1990). doi:10.1038/347539a0
12. J. Liu, L.N. Lewis, A.R. Duggal, Photoactivated and patternable charge transport materials and their use in organic light-emitting devices. Appl. Phys. Lett. **90**, 233503 (2007) doi:10.1063/1.2746404
13. M. Granstrom, K. Petritsch, A.C. Arias, A. Lux, M.R. Andersson, R.H. Friend, Laminated fabrication of polymeric photovoltaic diodes. Nature **395**(6699), 257–260 (1998). doi:10.1038/26183
14. P. Peumans, S. Uchida, S.R. Forrest, Efficient bulk heterojunction photovoltaic cells using small-molecular-weight organic thin films. Nature **425**(6954), 158–162 (2003). doi:10.1038/nature01949
15. P. Peumans, S.R. Forrest, Very-high-efficiency double-heterostructure copper phthalocyanine/C_{60} photovoltaic cells. Appl. Phys. Lett. **79**, 126 (2001). doi:10.1063/1.1384001
16. L. Torsi, A. Dodabalapur, L. Sabbatini, P. Zambonin, Multi-parameter gas sensors based on organic thin-film-transistors. Sens. Actuators, B **67**(3), 312–316 (2000). doi:10.1016/S0925-4005(00)00541--4
17. M. Shtein, J. Mapel, J.B. Benziger, S.R. Forrest, Effects of film morphology and gate dielectric surface preparation on the electrical characteristics of organic-vapor-phase-deposited pentacene thin-film transistors. Appl. Phys. Lett. **81**, 268 (2002). doi:10.1063/1.1491009
18. J.R. Heath, M.A. Ratner, Molecular electronics. Phys. Today **56**, 43 (2003). doi:10.1063/1.1583533
19. K. Suto, S. Yoshimoto, K. Itaya, Two-dimensional self-organization of phthalocyanine and porphyrin: dependence on the crystallographic orientation of Au. J. Am. Chem. Soc. **125**(49), 14976–14977 (2003). doi:10.1021/ja038857u
20. D. Heim, D. Écija, K. Seufert, W. Auwärter, C. Aurisicchio, C. Fabbro, D. Bonifazi, J.V. Barth, Self-assembly of flexible one-dimensional coordination polymers on metal surfaces. J. Am. Chem. Soc. **132**(19), 6783–6790 (2010). doi:10.1021/ja1010527
21. F. Buchner, I. Kellner, W. Hieringer, A. Görling, H. Steinrück, H. Marbach, Ordering aspects and intramolecular conformation of tetraphenylporphyrins on Ag(111). Phys. Chem. Chem. Phys. **12**(40), 13082–13090 (2010). doi:10.1039/C004551A
22. S. Weigelt, C. Busse, C. Bombis, M.M. Knudsen, K.V. Gothelf, T. Strunskus, C. Wöll, M. Dahlbom, B. Hammer, E. Lægsgaard, F. Besenbacher, T.R. Linderoth, Covalent interlinking of an aldehyde and an amine on a Au(111) surface in ultrahigh vacuum. Angew. Chem. Int. Ed. **46**(48), 9227–9230 (2007). doi:10.1002/anie.200702859
23. C. Wäckerlin, D. Chylarecka, A. Kleibert, K. Müller, C. Iacovita, F. Nolting, T.A. Jung, N. Ballav, Controlling spins in adsorbed molecules by a chemical switch. Nat. Commun. **1**, 61 (2010). doi:10.1038/ncomms1057
24. T. Ikeda, O. Tsutsumi, Optical switching and image storage by means of azobenzene Liquid-Crystal films. Science **268**(5219), 1873–1875 (1995). doi:10.1126/science.268.5219.1873
25. J. Henzl, M. Mehlhorn, H. Gawronski, K. Rieder, K. Morgenstern, Reversible cis-trans isomerization of a single azobenzene molecule. Angew. Chem. Int. Ed. **45**(4), 603–606 (2006). doi:10.1002/anie.200502229

26. L. Grill, M. Dyer, L. Lafferentz, M. Persson, M.V. Peters, S. Hecht, Nano-architectures by covalent assembly of molecular building blocks. Nat. Nanotechnol. **2**, 687–691 (2007). doi:10.1038/nnano.2007.346
27. A. Aviram, M.A. Ratner, Molecular rectifiers. Chem. Phys. Lett. **29**(2), 277–283 (1974). doi:10.1016/0009-2614(74)85031--1
28. C. Kergueris, J.-P. Bourgoin, S. Palacin, D. Esteve, C. Urbina, M. Magoga, C. Joachim, Electron transport through a metal-molecule-metal junction. Phys. Rev. B **59**(19), 12505–12513 (1999). doi:10.1103/PhysRevB.59.12505
29. A. Nitzan, M.A. Ratner, Electron transport in molecular wire junctions. Science **300**(5624), 1384–1389 (2003)
30. N.J. Tao, Electron transport in molecular junctions. Nat. Nanotechnol. **1**(3), 173–181 (2006). doi:10.1038/nnano.2006.130
31. J. Park, A.N. Pasupathy, J.I. Goldsmith, C. Chang, Y. Yaish, J.R. Petta, M. Rinkoski, J.P. Sethna, P.L. McEuen, D.C. Ralph, Coulomb blockade and the Kondo effect in single-atom transistors. Nature **417**(June), 722 (2002)
32. S.M. Lindsay, M.A. Ratner, Molecular transport junctions: clearing mists. Adv. Mater. **19**(1), 23–31 (2007). doi:10.1002/adma.200601140
33. L.H. Yu, D. Natelson, Transport in single-molecule transistors: kondo physics and negative differential resistance. Nanotechnol. **15**(10), S517 (2004). doi:10.1088/0957-4484/15/10/004
34. A.A. Houck, J. Labaziewicz, E.K. Chan, J.A. Folk, I.L. Chuang, Kondo effect in electromigrated gold break junctions. Nano Lett. **5**(9), 1685–1688 (2005). doi:10.1021/nl050799i
35. M.N. Baibich, J.M. Broto, A. Fert, F.N. Van Dau, F. Petroff, P. Etienne, G. Creuzet, A. Friederich, J. Chazelas, Giant magnetoresistance of (001)Fe/(001)Cr magnetic superlattices. Phys. Rev. Lett. **61**(21), 2472–2475 (1988). doi:10.1103/PhysRevLett.61.2472
36. G. Binasch, P. Grünberg, F. Saurenbach, W. Zinn, Enhanced magnetoresistance in layered magnetic structures with antiferromagnetic interlayer exchange. Phys. Rev. B **39**(7), 4828–4830 (1989). doi:10.1103/PhysRevB.39.4828
37. S.A. Wolf, D.D. Awschalom, R.A. Buhrman, J.M. Daughton, S.v. Molnár, M.L. Roukes, A.Y. Chtchelkanova, D.M. Treger, Spintronics: a spin-based electronics vision for the future. Science **294**(5546), 1488–1495 (2001). doi:10.1126/science.1065389
38. S. Sanvito, Molecular spintronics. Chem. Soc. Rev. **40**(6), 3336 (2011). doi:10.1039/c1cs15047b
39. R. Jain, K. Kabir, J.B. Gilroy, K.A.R. Mitchell, K.-c. Wong, R.G. Hicks, High-temperature metal-organic magnets. Nature **445**(January), 291–294 (2007)
40. W. Liang, M.P. Shores, M. Bockrath, J.R. Long, H. Park, Kondo resonance in a single-molecule transistor. Nature **417**(June), 725–729 (2002). doi:10.1038/nature00790
41. J.J. Parks, A.R. Champagne, T.A. Costi, W.W. Shum, A.N. Pasupathy, E. Neuscamman, S. Flores-Torres, P.S. Cornaglia, A.A. Aligia, C.A. Balseiro, G.K. Chan, H.D. Abruña, D.C. Ralph, Mechanical control of spin states in spin-1 molecules and the underscreened Kondo effect. Science **328**(5984), 1370–1373 (2010). doi:10.1126/science.1186874
42. N.B. McKeown, *Phthalocyanine Materials: Synthesis, Structure, and Function*, (Cambridge University Press, Cambridge, 1998), ISBN 9780521496230
43. J. Blochwitz, M. Pfeiffer, T. Fritz, K. Leo, Low voltage organic light emitting diodes featuring doped phthalocyanine as hole transport material. Appl. Phys. Lett. **73**(6), 729–731 (1998). doi:10.1063/1.121982
44. Y. Qiu, Y. Gao, P. Wei, L. Wang, Organic light-emitting diodes with improved hole-electron balance by using copper phthalocyanine/aromatic diamine multiple quantum wells. Appl. Phys. Lett. **80**(15), 2628–2630 (2002). doi:10.1063/1.1468894
45. S. Uchida, J. Xue, B.P. Rand, S.R. Forrest, Organic small molecule solar cells with a homogeneously mixed copper phthalocyanine: C_{60} active layer. Appl. Phys. Lett. **84**(21), 4218–4220 (2004). doi:10.1063/1.1755833
46. T. Kume, S. Hayashi, H. Ohkuma, K. Yamamoto, Enhancement of photoelectric conversion efficiency in copper phthalocyanine solar cell: white light excitation of surface plasmon polaritons. Jpn. J. Appl. Phys. **34**(Part 1, 12A), 6448–6451 (1995). doi:10.1143/JJAP.34.6448

References

47. C.J. Chen, *Introduction to Scanning Tunneling Microscopy*, (Oxford University Press, New York, 1993), ISBN 0195071506 9780195071504. http://www.columbia.edu/jcc2161/documents/STM_2ed.pdf
48. J. Gimzewski, E. Stoll, R. Schlittler, Scanning tunneling microscopy of individual molecules of copper phthalocyanine adsorbed on polycrystalline silver surfaces. Surf. Sci. **181**(1–2), 267–277 (1987). doi:10.1016/0039-6028(87)90167--1
49. H. Ohtani, R.J. Wilson, S. Chiang, C.M. Mate, Scanning tunneling microscopy observations of benzene molecules on the Rh(111)-(3 × 3) (C_6H_6 + 2CO) surface. Phys. Rev. Lett. **60**(23), 2398–2401 (1988). doi:10.1103/PhysRevLett.60.2398
50. P.H. Lippel, R.J. Wilson, M.D. Miller, C. Wöll, S. Chiang, High-Resolution imaging of copper-phthalocyanine by scanning-tunneling microscopy. Phys. Rev. Lett. **62**(2), 171–174 (1989). doi:10.1103/PhysRevLett.62.171
51. F. Moresco, Manipulation of large molecules by low-temperature STM: model systems for molecular electronics. Phys. Rep. **399**(4), 175–225 (2004). doi:16/j.physrep.2004.08.001

Chapter 2
Experimental Techniques

In this chapter we briefly introduce the experimental method—Scanning Tunneling Microscopy—used over the course of this thesis. First we review the main concepts and their theoretical background. Then we give a quick overview of the main experimental techniques like scanning tunneling spectroscopy and manipulation, followed by a short description of the actual setup and the sample preparations used throughout this thesis.

2.1 Scanning Tunneling Microscopy

Scanning Tunneling Microscopy (STM) is an experimental technique in surface physics that allows direct imaging of conducting and semiconducting surfaces on the nanometer scale (Fig. 2.1). By using a local probe, images are acquired in real space complementing other techniques like Low Energy Electron Diffraction (LEED) that work in reciprocal space. Additionally, and more importantly perhaps, the STM provides spatially resolved access to the electronic structure of the sample. Since its development in 1981 [1–3], STM has become a powerful and widely used tool in surface science [4]. Further improvements like low temperature setups allowed the investigation of systems of absorbates, where a high mobility at room temperature would have eluded scanning, such as single molecules or adatoms. Especially these kind of setups promoted the use of STM not only as an imaging tool but also as means to manipulate the sample in a very precise way, for instance by creating nano-sized structures by moving single atoms and molecules [5, 6] or inducing localized chemical reactions [7].

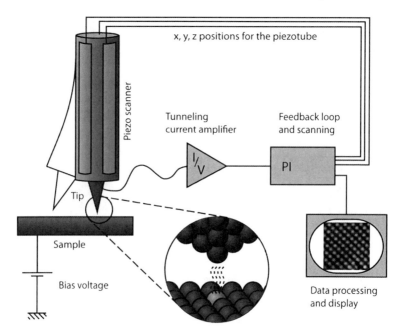

Fig. 2.1 The scanning tunneling microscope: a sharp conducting tip is approached very close (~5Å) to the sample surface. Due to the bias voltage between tip and sample a current of tunneling electrons can be detected, which depends on the local characteristics of the sample at the position of the tip. Using a piezo electronic actuator the tip is moved in a scanning motion over the sample surface. A topographic image of the surface can be constructed based on the tunneling current signal

2.1.1 The Operating Principle

The basic idea of an STM is based on quantum tunneling. Consider a particle with a given energy that is confronted with a barrier in its trajectory. In classical mechanics it cannot cross the barrier if its potential is higher than its own energy. An example of this would be a ball trying to roll up a hill. However, if one looks at a system that is small enough to require a quantum mechanical description, the outcome of this experiment will be different. In quantum mechanics matter is seen as a wave and a particle at the same time. The position and the momentum of an object can only be determined with a certain degree of uncertainty. If the momentum is very sharply defined the position cannot be determined a vice versa, according to the Heisenberg uncertainty principle. This implies that when a quantum mechanical particle faces a potential energy barrier, the probability that the particle crosses the barrier is not zero, but instead decays exponentially over the width of the barrier. Hence, there will always be a small probability for the particle to emerge on the other side of the barrier (see Fig. 2.3). Naturally this probability will depend on the height and width of the barrier.

2.1 Scanning Tunneling Microscopy

In an STM this effect is harnessed by approaching a very sharp metallic tip, ideally terminated by just one atom, at a distance of about 0.1–1 nm to a conducting or at least semiconducting sample. The gap between them is the barrier that electrons will tunnel though. In the absence of a bias, both directions for the tunneling from tip to the sample and vice versa are equally probable, and hence no net electron current would be detectable. But when a bias voltage between the sample and the tip is applied one tunneling direction is favored and a measurable current is detected. This tunneling current (typically 1 pA to 10 nA) can be measured and depends exponentially on the distance between the sample and the tip.

Now if the tip scans an area on the surface at a constant height, the current signal is an image of the topography in real space. This is called the constant height scanning mode (see Fig. 2.2). There are some disadvantages to this scanning mode. If the surface has topographic features higher than the scan height, the tip will crash against them, thus changing the surface and the tip. Further, the vertical height of the tip has to be very stable with respect to thermal drift or coupling to external vibrations.

Another more commonly used scanning mode is the constant current mode (see Fig. 2.2). In this case a feedback loop is used to maintain the current constant, by varying the height of the tip, i.e., the tunneling barrier. In this constant current mode the topographic information is recorded in the height control signal during the scan. Because the tip follows the profile of the surface, this mode is better suited for bigger corrugations. In this work the constant current mode was used almost exclusively.

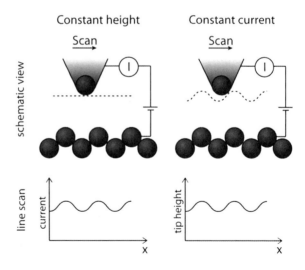

Fig. 2.2 The different STM scanning modes: In constant height mode the tip is scanning at a constant height, the tunneling current signal contains the topographic data. In constant current mode the tunneling current is kept constant by a feedback loop controlling the tip height, the topographic signal is the tip height

2.1.2 Theoretical Descriptions of the Tunneling Process

The tunneling process is described by a variety of different theoretical models. Two of which will be discussed in brief here. The first is a simple time-independent 1D model. Despite its approximative character, this model is very instructive because it can be solved analytically and leads to the fundamental dependencies of the tunneling current. The second model takes all three dimensions into account, as well as the electronic structure of tip and sample.

Simple 1D Model

In a first approach the tunnel junction can be modeled as a one dimensional and time independent system. The electron is approximated as a free electron with energy E, separated by a potential barrier of height Φ and thickness d.

The stationary Schrödinger equation (Eq. 2.1) for the electron wave functions Ψ in the tip, the sample, and the barrier has to be solved:

$$\left(\frac{\hbar^2}{2m_e}\Delta + V(r)\right)\Psi = E\Psi \qquad (2.1)$$

where m_e is the electron mass, and $V(r)$ the potential describing the barrier with height Φ and width d and E the energy of the electron. The exact solution for these wave functions can be found by using a plane wave ansatz for Ψ for the three regions (tip (1), vacuum (2) and sample (3), see Fig. 2.3).

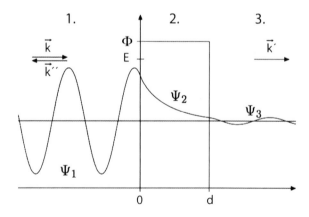

Fig. 2.3 One-dimensional potential barrier with energy Φ and width d. A particle that is traveling from the left to the right is described by three wavefunctions Ψ before (*1*), inside (*2*), and after (*3*) the barrier. This leads to three wavenumbers $\vec{k}, \vec{k}'$ and $\vec{k}''$. The $\vec{k}'$ solution crossed the barrier

2.1 Scanning Tunneling Microscopy

$$\Psi_1 = e^{ikz} + A \cdot e^{-ikz}$$
$$\Psi_2 = B \cdot e^{i\kappa z} + C \cdot e^{-i\kappa z}$$
$$\Psi_3 = D \cdot e^{ikz}$$

$$\text{with } k = \frac{\sqrt{2m_e E}}{\hbar} \text{ and } \kappa = \frac{\sqrt{2m_e(\Phi - E)}}{\hbar}$$

By matching the amplitude and the first derivative at the boundaries of the different potential regions, the coefficients (A, B, C, D) can be determined [8].

We can then define a transmission coefficient T by comparing the wavefunctions on both sides of the barrier:

$$T = \frac{|\Psi_1|^2}{|\Psi_3|^2} = \frac{A^2}{D^2} = \left[\left(\frac{k^2 + \kappa^2}{2k\kappa}\right)^2 \sinh(\kappa d)\right]^{-1} \quad (2.2)$$

which can be simplified for a high barrier potential compared to the energy of the electron $\Phi \gg E$, i.e., if $\kappa d \gg 1$

$$T \approx \frac{16k^2\kappa^2}{(k^2 + \kappa^2)^2} \cdot e^{-2\kappa d} \quad (2.3)$$

The number of tunneling electrons, that is the tunneling current I, will be proportional to T. Therefore with the thickness of the barrier given by the tip height d, we find:

$$I \propto T \propto e^{-2\kappa d} \quad (2.4)$$

We have thus shown that the tunneling current depends exponentially on the tip-sample distance d. The barrier height Φ is the average work function of the sample and tip. For typical values of $\Phi = 5$ eV and $E = 2$ eV, this means a change of one order of magnitude for I for change of 1 Å in the tip sample distance d. This exponential dependency is the underlying mechanism for the high vertical resolution of the STM technique.

The Tersoff-Hamann Model

This model can be extended by including the electronic structure of the tip and sample and the three dimensions of the problem. Tersoff and Hamann presented this more realistic model in the 1980s [9, 10]. Their starting point was a first-order perturbation theory model introduced by Bardeen [11]. In this model the tunneling is described by the tunneling matrix $M_{\mu,\nu}$, which represents the overlap of the wave functions of tip Ψ_μ and surface Ψ_ν.

The tunneling current for a general geometry can then be written as:

$$I = \frac{2\pi e}{\hbar} \sum_{\mu,\nu} f(E_\mu)[1 - f(E_\nu + eV)]|M_{\mu,\nu}|^2 \delta(E_\mu - E_\nu) \quad (2.5)$$

Where $f(E)$ is the Fermi-Dirac distribution function and E_μ and E_ν the energies of the states of surface and tip. V is the applied bias voltage. In the limit of low temperature and voltages $e \cdot V \ll \Phi$, Eq. 2.5 can be simplified because $f(E)$ becomes a step function with $f(E) = 1$ for $E < E_F$. We can thus approximate the term $[1 - f(E_\nu + eV)]$ as a step function, which yields:

$$I = \frac{2\pi}{\hbar} e^2 V \sum_{\mu,\nu} |M_{\mu,\nu}|^2 \delta(E_\mu - E_F) \delta(E_\nu - E_F) \quad (2.6)$$

The tunneling matrix element $M_{\mu,\nu}$, was shown by Bardeen [11] to be Eq. 2.7 integrated over an arbitrary surface between tip and surface:

$$M_{\mu,\nu} = -\frac{\hbar^2}{2m} \int \left(\Psi_\mu^* \vec{\nabla} \Psi_\nu - \Psi_\nu \vec{\nabla} \Psi_\mu^* \right) d\vec{S} \quad (2.7)$$

In order to arrive at quantitative tunneling currents it is necessary to define the wave functions for surface and tip. These should describe the geometry found in an STM. The surface is described through a wave function parallel to the surface, consistent with Bloch's theorem. In the perpendicular direction it decays exponentially into the vacuum:

$$\Psi_\nu = V_{surface}^{-1/2} \sum_G a_G \underbrace{e^{\left(-\sqrt{\kappa^2 + |\vec{k}_\| + \vec{G}|^2}\, z\right)}}_{\text{Exp.decay}} \cdot \underbrace{e^{\left(i[\vec{k}_\| + \vec{G}]\vec{x}\right)}}_{\text{Blochwave}} \quad (2.8)$$

where $\vec{G}$ is a reciprocal lattice vector, $\kappa = \hbar^{-1}(2m\Phi)$ the decay length into the vacuum, $\vec{k}_\|$ the wave vector of the surface wave, and $V_{surface}$ the normalization volume of the surface. The first few coefficients a_G are typically of order unity. The tip is modeled as a spherical potential (see Fig. 2.4) at the point closest to the surface and the rest is arbitrary. Accordingly the spherical s-wave function is used for the tip:

$$\Psi_\mu = V_{tip}^{-1/2} \kappa R\, e^{\kappa R} \frac{1}{\kappa |\vec{r} - \vec{r}_0|} e^{-\kappa |\vec{r} - \vec{r}_0|} \quad (2.9)$$

here V_{tip} is the normalization volume of the tip, R the tip radius, κ and Φ are the same constants as mentioned above. For simplicity's sake the work function of both tip and surface are assumed to be equal. With these model wave functions, it is possible to simplify the tunneling matrix (Eq. 2.7) to:

$$M_{\mu,\nu} = \frac{\hbar^2}{2m} 4\pi k^{-1} V_{tip}^{-1/2} k R e^{kR} \Psi_\nu(\vec{r}_0) \quad (2.10)$$

2.1 Scanning Tunneling Microscopy

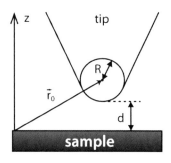

Fig. 2.4 Model used to approximate the wave function of the tip in the Tersoff Hamann description. The center of the tip is at $\vec{r}_0$ with an assumed spherical geometry, R is the radius and d the distance to the surface

Finally, Eq. 2.10 can be used to define the tunneling current (Eq. 2.6) giving:

$$I = 32\frac{\pi^3}{\hbar k^4}e^2 V \Phi^2 R^2 e^{2kR} \frac{1}{V_{tip}} \sum_{\mu\nu} |\Psi_\nu(\vec{r}_0)|^2 \delta(E_\mu - E_F)\delta(E_\nu - E_F) \quad (2.11)$$

To simplify this expression we note that the local density of states for tip and surface are defined as follows:

$$\rho_{tip}(E) = \frac{1}{V_{tip}} \sum_\mu \delta(E_\mu - E) \quad (2.12)$$

$$\rho_{surface}(E, \vec{r}_0) = \sum_\nu |\Psi_\nu(\vec{r}_0)|^2 \delta(E_\nu - E) \quad (2.13)$$

The final expression for the tunneling current comes to:

$$I \propto V\rho_{tip}(E_F)\rho_{surface}(E_F, \vec{r}_0) \quad (2.14)$$

Thus the current depends on the local density of states (LDOS) of the surface at the position of the tip $\vec{r}_0$ at the Fermi-Energy E_F. This means that STM images the LDOS of the surface rather then the position of atoms.

A more realistic description, a generalization of the Tersoff-Hamann model, defines the tunneling current by integrating over the states contributing to the tunneling current: the surface and tip DOS within the finite bias window. The dependence of the vacuum barrier on the tip-sample distance d, the energy E of each state, and the bias voltage V are represented by a transmission coefficient $T(d, E, eV)$ [12–14].

$$I \propto \int_{E_F}^{E_F+eV} \rho_{surface}(E_F - eV + \varepsilon)\rho_{tip}(E_F + \varepsilon)T(d, \varepsilon, eV)d\varepsilon \quad (2.15)$$

We have thus found the contributing factors to the tunneling current: The DOS of tip and sample, and a transmission coefficient $T(d, E, eV)$. For small biases

ρ_{tip} and T can be assumed constant; the tunneling current would then be primarily proportional to $\rho_{surface}$ integrated from E_F to the applied bias voltage $E_F + V$.

This model is still not a complete description of the tunneling process in an STM. Although in many standard situations it provides a reasonable qualitative picture. Some critical remarks have to be added nonetheless. The approximation of the tip as an s orbital is rather inaccurate, as in tungsten tips the d orbitals contribute to the major part of the tunneling current [15]. Also, the interaction between tip and sample through microscopic chemical forces is not always negligible.

2.2 Scanning Tunneling Spectroscopy

2.2.1 Elastic Tunneling Spectroscopy

One of the key features of an STM setup is the ability to measure the electronic structure of a sample at the position of the tip. The basic principles of the scanning tunneling spectroscopy (STS) mode can be understood from Eq. 2.15. If the transmission coefficient T and to some extent ρ_{tip} are assumed to be constant for the given energy interval, we can approximate the differential conductance dI/dV as:

$$\frac{dI}{dV} \propto \rho_{surface}(E_F - eV) \cdot \rho_{tip}(E_F) \qquad (2.16)$$

The differential conductance dI/dV is hence directly proportional to the local density of states of the sample at the position of the tip. Depending on the sign of applied bias the occupied or the unoccupied states are measured (see Fig. 2.5).

In STM experiments the STS signal is obtained by positioning the tip over the point of interest, then the tunneling gap is set to an appropriate setpoint, the feedback loop is opened to allow the current to vary and the voltage is ramped to cover the region of interest. The voltage setpoint should be set value that integrates the same DOS in all points investigated on the sample, to avoid strong variations in the tip height. This is, however, a complex task in inhomogeneous systems such as molecular adsorbates on surfaces, and not always possible due to limitations in the measurement apparatus. Hence, usually, while not entirely correct, the starting voltage of the voltage ramp was used in this work.

2.2.2 Inelastic Tunneling Spectroscopy

In addition to the elastic tunneling, electrons can couple to internal degrees of freedom during the tunneling process. Examples of these are molecular vibrations [7], and spin excitations [16]. In the case of tunneling to/from a molecule, inelastic processes

2.2 Scanning Tunneling Spectroscopy

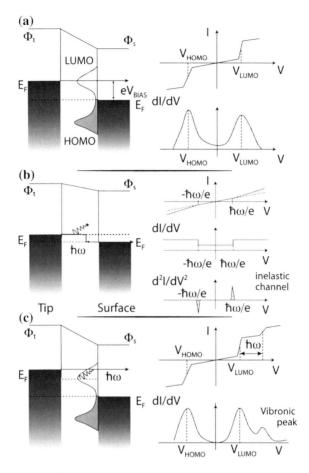

Fig. 2.5 Schematics of STS through a molecule. Energy level diagrams showing elastic and inelastic processes, occurring at positive biases. Φ_t and Φ_s are the work functions of tip and sample, respectively. **a** Elastic tunneling: electrons tunnel from the tip into the sample probing the unoccupied states around E_F of the sample. The dI/dV curve shows a peak at the position of a molecular orbital. **b** Inelastic tunneling. Electrons excite a vibration in the molecule and lose energy. The opening of this inelastic channel increases the conductance at the energy of the vibration. This is resolved as symmetric steps in the dI/dV curve and antisymmetric peaks at opposite polarities in the d^2I/dV^2 spectrum. **c** Tunneling process with vibronic excitation at a molecular orbital. The vibronic peak is represented as extra protrusions in the dI/dV spectra at energy separations of the excited phonon

can be excited, depending on the cross-section between molecular orbital and the excitation, the orientation of the molecule, the position of the tip, etc. Once an electron has the required energy to excite the process, the interaction leads to the opening of another tunneling channel. Such processes lead to a change in the slope of the I/V curve, that is, to a symmetric step function in the dI/dV and antisymmetric peaks in the d^2I/dV^2 (Fig. 2.5b). By recording the d^2I/dV^2 curve these excitations can be effectively measured. Resonant tunneling through electronic states can also couple to inelastic processes. If the lifetime of a given electronic state is long enough, such as in the case of decoupled molecular orbitals, an electron tunneling to this state can couple to vibrations. These so-called vibronic excitations are then detected as extra satellite peaks in the dI/dV signal at multiples of the energy of the vibrational mode above the molecular resonances (Fig. 2.5c).

2.2.3 Lock-in Technique

In an experimental setup the dI/dV signal can be recorded directly using a lock-in amplifier. Compared to a numerical calculation from the I/V curve, this method yields a better dynamic range and signal to noise ratio.

A small sinusoidal voltage modulation ($V_{mod} = v_o \sin(\omega_r t)$, with $v_o = 0.1$ to 100 mV) is added to the bias voltage, causing a sinusoidal response in the tunneling current. The amplitude of the modulated current is sensitive to the slope of the I-V curve (see Fig. 2.6). The resulting current signal is then converted by the preamplifier to a voltage:

$$V_{in} = V^I_{sig} \sin(\omega_r t + \theta_{sig}) + V^I_{DC} + V^I_{noise}(t) \qquad (2.17)$$

where V^I_{DC} is the DC part of the current, and V^I_{noise} the noise present in the system θ_{sig} is the phase of the measurement signal. The lock-in then multiplies this signal with a sinus wave $V_{ref} = V_l \sin(\omega_r t + \theta_l)$, resulting in an output of:

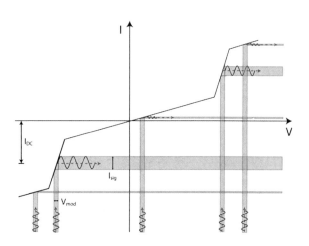

Fig. 2.6 The origin of the derivative in a lock-in measurement. During an I-V ramp for STS a sinusoidal signal is added to the bias voltage. The resulting modulation in the tunneling current has an amplitude proportional to the slope of the I-V curve, hence a dI/dV signal

2.2 Scanning Tunneling Spectroscopy

$$\begin{aligned}V_{psd} &= V_{sig} V_l \sin(\omega_r t + \theta_{sig}) \sin(\omega_l t + \theta_l) \\ &\quad + V_{ref}(t) V_{DC} + V_{ref}(t) V_{noise}(t) \quad (2.18)\\ &= 1/2 V_{sig} V_{ref} \cos([\omega_r - \omega_r]t + \theta_{sig} - \theta_l) \\ &\quad - 1/2 V_{sig} V_{ref} \cos([\omega_r + \omega_r]t + \theta_{sig} + \theta_l) \\ &\quad + V_{ref}(t) V_{DC} + V_{ref}(t) V_{noise}(t) \quad (2.19)\end{aligned}$$

The signal is then passed through a low pass filter, which is done integrating the signal over a given time τ. A higher integrating time means a lower pass bandwidth, resulting in a better signal-to-noise ratio. The time dependent terms will be removed by this operation, leaving only the first term of Eq. 2.19:

$$V_{psd} = 1/2 V_{sig} V_l \cos(\theta_{sig} - \theta_{ref}) \quad (2.20)$$

and only the Fourier components of the noise term with frequencies close to the reference ω_r. This way any noise present in the signal will be strongly reduced by the band pass filter. To minimize the Fourier components of the noise passing the filter, frequencies which do not interfere with the internal resonances (mechanical or electrical) of the experimental setup and are within the bandwidth of the current amplifier are used ($\omega_r = 0.5-3$ kHz).

The lock-in electronics assure that the phase relation ($\theta_{sig} - \theta_{ref}$) between the lock-in modulation and the measurement signal is time-independent. Its value should be chosen so that the detected signal becomes maximum $\cos(\theta) = 1$.

Higher derivatives can also be measured using higher harmonics of the signal [17]. The second harmonic will be used to perform inelastic spectroscopy via the d^2I/dV^2 signal.

The dI/dV signal measured in units of conductance [nS = nA/V/V]. The conversion between the lock-in output and conductance units follows directly:

$$\frac{dI}{dV} = \left(\text{signal} \cdot \frac{\text{lock-in sensitivity}}{\text{lock-in full scale}} \cdot \text{preamplifier gain}\right) \cdot \frac{1}{\text{lock-in modulation}}$$

2.2.4 Background Subtraction

The dI/dV signal is convoluted with the DOS of the tip, which in real experiments is *not* constant. Therefore great care has to be taken to prepare the tip, not only to be sharp in order to achieve good topography resolution, but also to obtain a featureless DOS. However, the shape and DOS of the tip can only be manipulated in very indirect ways. A wide range of methods have been developed, e.g. by voltage pulses, or small controlled contacts with the surface, but non of them is very reproducible and they all

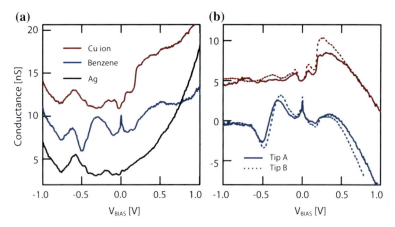

Fig. 2.7 **a** dI/dV spectra acquired on the Cu ion (*red*), and benzene (*blue*) of a single CuPc molecule on a Ag(100) surface, and on a clean spot of the Ag (*black*) with tip A (3.0 nA, −1.0 V). The spectra have been y-offset for clarity, all of them start at ~10 nS, like the blue spectra. **b** The same ion and benzene spectra after subtraction of the Ag spectrum (*solid lines*). For comparison, we show a pair of background subtracted spectra (*dotted lines*) acquired with a different tip (B) and initial setpoint (3.0 nA, −2.0 V). The intensity of the latter have been reduced by a factor of 2 to account for the different initial setpoint. The spectra recorded on the Cu ion have been vertically shifted by 5 nS for better visibility

are time intensive. The results will vary from tip to tip, and the tip may even change in between STS measurements.

A very efficient method to reduce the impact of tip DOS features is background subtraction. The basic idea is to measure the LDOS of the tip and the surface on a clean spot of the sample surface using the same feedback conditions as for the sample's feature of interest. This data can then be used to deconvolute effects on the STS taken on the point of interest. Theoretically a mathematical deconvolution operation would be required to remove tip and surface contribution of the STS. However, as P. Wahl et al. pointed out [18], a subtraction of the two STS curves is often sufficient, and leads to good results if the tip reference spectra does not present any sharp features. Figure 2.7 shows an example of background subtraction of dI/dV spectra recorded on CuPc/Ag(100). It can be observed that after subtraction the spectra taken with different tips are essentially the same. Furthermore, the tip related feature appearing around −0.6 V in the raw spectrum is removed, showing the efficacy of this method. Throughout this work all background subtracted spectra are obtained in this manner.

Nonetheless there are certain artifacts that can be created by this method, i.e., strong tip features can lead to artificial peaks in the resulting curves. Furthermore the conductance values reported for the spectra represent relative values. Negative dI/dV values mean that the conductance at the point of interest at a certain energy is lower than that of the substrate (see Fig. 2.7b).

2.2.5 Differential Conductance Maps

An intriguing application of STM lies in the possibility to spatially resolve the topography of subnanoscale systems. The technique can be combined with STS, mapping the spatial distribution of features found in STS. The basic method is to scan an area of interest and take the dI/dV intensity at each point of the scan. The resulting dI/dV map is to some extent an image of the spatial distribution of the state investigated. The speed of such an acquisition is limited by the τ setting of the lock-in amplifier (5–8 τ per point) and the pixel resolution of the dI/dV map. Depending on its size the measurement time in our experiments varied between 5–30 min, meaning that the effect of thermal drift becomes important. To counteract the thermal drift in the z direction the constant current mode is usually employed.[1]

The interpretation of such dI/dV maps is however not straightforward, because the topography of the sample has a strong influence on the height of the tip and will enhance or reduce certain spectral features. A variety of normalization techniques for dI/dV spectra and maps exist, and provide useful results in particular situations [14, 19–22]. In any case, the qualitative comparison of molecular states for similar molecules in different environments, as presented in later chapters, is not severely affected by topographic effects, therefore in this work no normalization techniques were used.

2.3 Manipulation Techniques

Another powerful feature of STM at low temperature is the possibility to locally manipulate the investigated system, by moving atoms or molecules with the tip. This is a great tool for scientific research as it allows control of the positions of single atoms and molecules to create new nanosystems and investigate them. The manipulation process is controlled through three main parameters in the STM: the electric field, the tunneling current, and the short range interactions between tip and surface [6, 23]. Care has to be taken to find for each specific system the right combination of parameters. There are two main methods employed for manipulation: vertical and lateral manipulation (see Fig. 2.8).

In the vertical manipulation mode, field and current effects play a major role. Here the tip is also approached towards the adsorbate while a voltage pulse is applied. With the correct parameters this transfers the adsorbate to the tip, which is effectively "picked up". Afterwards the tip is retracted and moved to a new location where a reverse pulse and approach drops the adsorbate again. In a related application this method can be used to modify the tip, to gain access to different spectral features due to changes of the symmetry of tunneling orbitals [24].

In the lateral manipulation mode, the tip is approached towards the surface, on top of the atom/molecule that has to be moved, by increasing the current and lowering

---

[1] However with carefully chosen parameters a constant height map is also possible.

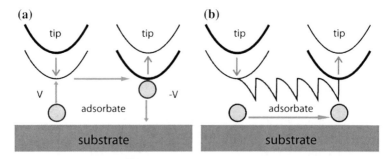

Fig. 2.8 a Sketch of the vertical manipulation procedure: 1. The tip is placed above the molecule/atom 2. To transfer an adsorbate to the tip a voltage pulse is applied while moving the tip towards the atom 3. The tip is retracted and moved to the desired position 4. The reverse pulse from the pick up is applied to drop the adsorbate. **b** The lateral manipulation process in constant current mode: 1. The tip is placed above the molecule/atom 2. The tip is approached towards the atom by increasing the current and decreasing the bias voltage 3. The tip is moved along the desired pathway. The tip height curve shows the movement of the adsorbate on the surface. Images adapted from [6]

the bias voltage (e.g. for Li on Ag(100) the conditions were 50 mV, 250 nA). Then the tip follows the desired pathway and finally is retracted again. If the interaction between tip and atom, such as vdW, chemical forces or the electric field, is strong enough the atom is moved along with the tip. The molecules can either be pushed or pulled depending on the specific system, leading to characteristic tip height curves. A pulling movement, consists of a step increase as the molecule is pulled under the tip, and subsequent slow decrease as the tip moves away from it (see Fig. 2.9a). For a pushing manipulation the curve would be mirrored, starting with a slow rise, followed by a sharp drop as the molecule jumps forward.

Both types of manipulation were used for the first time by the group of Don Eigler at IBM-Almaden [25] and later in Berlin [26] to build nanostructures from single atoms and molecules. Throughout this work mainly the lateral manipulation mode has been used to move atoms and molecules. In Fig. 2.9 the creation of a 3x3 cluster of CuPc molecules is shown.

2.4 Experimental Setup

The experimental setup used in this work is a low temperature STM based on a design by G. Meyer [27], and commercially available from Createc Fischer and Co GmbH Berlin, Germany [28]. The system operates in an ultra high vacuum (UHV) environment ($p < 2 \times 10^{-10}$ mBar), and is based on a liquid helium cryostat to work at temperatures ~5 K. In this section a brief review of the capabilities and limitations of this STM will be presented.

2.4 Experimental Setup 23

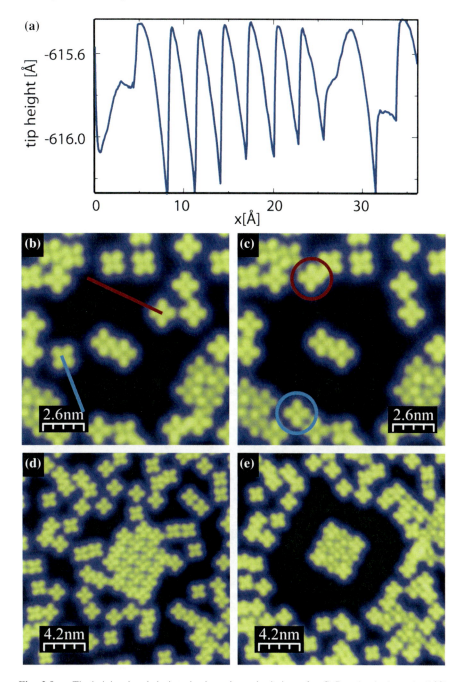

Fig. 2.9 **a** Tip height signal during the lateral manipulation of a CuPc adsorbed on Ag(100), corresponding to a pulling movement **b** and **c** STM topography before and after dragging two CuPc molecules. A gap resistance of 333 KΩ (1.5 × 10⁵ nA, 50 mV) was used for manipulation, and 1000 MΩ (0.1 nA, 100 mV) for imaging. **d** and **e** Images before and after the formation of a 3 × 3 square cluster (see Sect. 5.4.1)

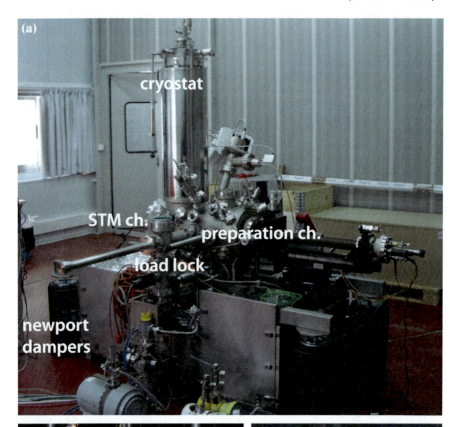

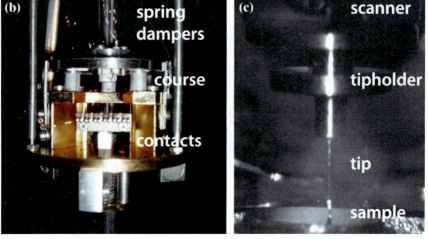

Fig. 2.10 **a** The low temperature STM vacuum chamber: The STM is located below the cryostat, the preparation chamber features a sputter gun for cleaning the samples, a LEED optic, and several evaporators: The quadruple OMBE for molecular sources, a triple Omicron for metals. The STM chamber has an in-situ Alkali evaporator and a single metal evaporator. **b** The low temperature STM head and the sample stage is shown: The sample holder is placed between the coarse motion ramp and the contacts. The whole setup is hanging on three springs to decouple it from external vibrations. **c** The piezo scanner and the tip in a tunneling position

Figure 2.10 shows the vacuum chamber of the STM: It consists of two chambers, one used for sample preparation and a second that houses the STM and its cryostat. The UHV is generated by an array of pumps: Two ion pumps one in each chamber, a turbo-molecular pump and a titanium sublimation pump on the preparation chamber. The cryostat itself also acts as cryo-pump in the STM chamber.

The Cooling System

The low operation temperatures are achieved by the use of a bath cryostat, made of two tanks. The inner one holds up to 11 litres of liquid helium while the outer one is filled with liquid nitrogen. Both cooling stages have aluminum radiation shields that enclose the STM sample stage within. Using this method the sample can be kept cool and measured for up to 100 h per refill of the helium deposit. The nitrogen tank lasts for approximately 48 h.

STM Head

The core of the STM setup is the STM head, shown in Fig. 2.10b. It hangs from the bottom of the He tank on three springs, which together with an Eddy current damping system decouple it from external vibrations. The STM head uses a Besocke type ramp to approach the tip to the sample until a tunneling current is detectable. The Besocke system is based on a plate that has three low angled ramps on its bottom side. This plate lies on three piezo tube actuators, which can displace the plate in x-y direction or rotate it, using a slip stick motion. Due to the ramps in the plate a rotation results in a z movement approaching or retracting the tip to/from the sample (maximum range 300 μm). The x-y movements of the ramp allow access to a large part of the sample surface for scanning. A fourth piezo tube actuator attached to the center of the ramp is the actual piezo scanner, which performs the x-y-z position during the STM measurements. The tip is placed in an in-situ exchangeable tip holder, which is attached to the piezo tube by a small magnet (see Fig. 2.10c).

Sample Preparation Facilities

The system is equipped with a sputter gun in the preparation chamber. To obtain atomically clean metal surfaces, the samples are bombarded with Ar+ ion, which removes impurities and adsorbates. The increased surface roughness is then smoothed by annealing the sample with an oven that is mounted on the sample holder. This process has to be repeated a number of times to obtain a good sample surface. The overall quality of the surface can be checked with a LEED setup, before and after deposition of molecules/atoms etc. on the surface. Several evaporator slots are present and a wide range of samples can be prepared: A commercial triple e-beam metal evaporator can be used to grow thin metal films, a quadruple Kundsen cell evaporator deposits molecules, and several single evaporators for molecules and metals are available. The main manipulator of the chamber can be cooled with liquid nitrogen, so that the sample temperature during deposition can be chosen between −180 °C to 800 °C. It is further possible to deposit materials in-situ on the sample, by a small door in the cooling shields of the STM. This slot is also equipped with a fixed home-built alkali evaporator, working with cells from SAES getters.

2.5 Methods

Sample Preparation

In order to investigate MePc (Me = Fe,Co,Ni,Cu) molecules adsorbed on the Ag(100) substrate, we prepared several samples of varying molecular coverages. To avoid contamination all experiments were undertaken under ultra high vacuum (UHV) conditions ($p < 2 \times 10^{-10}$ mbar). Before the MePc (Sigma Aldrich, 99% pure powder) they can be used in UHV, it is necessary to throughly degas the powder to achieve a base pressure during evaporation below $p < 5 \times 10^{-10}$ mbar.

First we cleaned the Ag(100) (Surface Preparation Laboratories) surface by repeated sputtering with Ag^+ ions and subsequent annealing to 450°C. We monitored the surface quality by Low Energy Electron Diffraction (LEED). Once a clean surface was obtained the molecules were evaporated from the molecular beam source onto the sample, which was kept at room temperature. The Knudsen cell of the evaporator was heated to 440°C, [2] which yielded a deposition rate of approximately 0.05 molecular layers per minute. In this way samples of different MePc coverages could be produced, which were then transferred into the STM sample stage. All measurements were carried out at 5 K in order to inhibit molecular mobility and perform single molecule spectroscopy and manipulation.

The doping experiments were prepared by in-situ deposition of Li or Fe atoms, as a second step after a MePc preparation. The source for the Li was a well degassed Li dispensor (SAES) heated by direct current flow (6 A). For the Fe atoms an electron beam evaporator was used.

STS Settings

The dI/dV spectra were obtained using the lock-in technique, using a bias voltage modulation of frequency around 3 kHz and amplitude 1 mV_{rms} for the low range spectra, 3 mV_{rms} for the larger energy range and 10 mV for dI/dV maps. Since zero bias resonances cannot be measured with closed feedback, we have mapped the conductance near E_F by measuring the second derivative of the current, mapping the maximum of the d^2I/dV^2 signal, which is slightly offset from $V_b = 0$, as shown in

---

[2] This temperature is measured on the outside of the crucible close to the filament and therefore does not represent be the actual temperature of the molecules during the deposition.

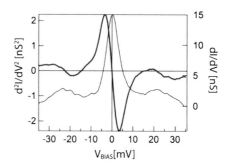

Fig. 2.11 Comparison between first (*black*) and second derivative (*blue*) of a CuPc spectrum showing a Kondo resonance. The maximum of the second derivative at negative voltage is used to map the Kondo resonance

Fig. 2.11. The validity of this approach is justified by the fact that, for a Lorentzian function, the intensity maximum of the peak is proportional to the (shifted) maximum of its derivative.

Density Functional Theory Methods

The spectroscopic results have been compared to the electronic structure obtained from ab-initio calculations performed by R. Robles, N. Lorente, R. Korytár, and P. Ordejón, using the VASP implementation of Density Functional Calculations (DFT) in the projected augmented plane wave scheme [29, 30]. Different approximations, each with different limitations and over- or underestimations have been compared in order to obtain consistent results, and to check the effect of different exchange-correlation approximations or the van der Waals (vdW) interaction. Namely the Local Density Approximation (LDA) [31], Generalized Gradient Approximation (GGA) [32], GGA+vdW [33, 34], the DSRLL functional, designed to include VdW [35–38], and LDA+U [39] were used. The plane wave cutoff energy was set to 300 eV. The calculated slab included 5 Ag atomic layers intercalated by 7 vacuum layers in the vertical direction, and a 7×7 lateral supercell. The positions of all atoms in the molecule and the first three Ag layers were relaxed vertically and laterally until forces were smaller than 0.05 eV/Å. The Projected Density of States (PDOS) of NiPc has been used to compare single ($1 \times 1 \times 1$) and multiple ($5 \times 5 \times 1$) k point calculations, which converge after applying a broadening of 100 meV to the data. Based on that, the results for a single k point with the latter broadening are used in the following. Charge transfer and local magnetic moments have been calculated using a Bader charge analysis [40, 41].

References

1. H. Binnig, G. Rohrer. Scanning tunnelling microscopy. Helv. Phys. Acta. **55**(6), 726–735 (1982)
2. G. Binnig, H. Rohrer, C. Gerber, E. Weibel, Tunneling through a controllable vacuum gap. Appl. Phys. Lett. **40**(2), 178–180 (1982)

3. G. Binnig, H. Rohrer, C. Gerber, E. Weibel. Surface studies by scanning tunneling microscopy. Phys. Rev. Lett. **49**(1), 57 (1982)
4. G. Binnig, K.H. Frank, H. Fuchs, N. Garcia, B. Reihl, H. Rohrer, F. Salvan, A.R. Williams. Tunneling spectroscopy and inverse photoemission: image and field states. Phys. Rev. Lett. **55**(9), 991 (1985)
5. H.C. Manoharan, C.P. Lutz, D.M. Eigler. Quantum mirages formed by coherent projection of electronic structure. Nature **403**(6769), 512–515 (2000). doi:10.1038/35000508
6. F. Moresco. Manipulation of large molecules by low-temperature STM: model systems for molecular electronics. Phys. Rep. **399**(4), 175–225 (2004). doi:16/j.physrep.2004.08.001
7. W. Ho. Single-molecule chemistry. J. Chem. Phys. **117**, 11033 (2002). doi:10.1063/1.1521153
8. C. Cohen-Tannoudji. Quantenmechanik 1. (Walter de Gruyter, Berlin, 1999), p. 64ff
9. J. Tersoff, D.R. Hamann. Theory and application for the scanning tunneling microscope. Phys. Rev. Lett. **50**(25), 1998 (1983)
10. J. Tersoff, D.R. Hamann. Theory of the scanning tunneling microscope. Phys. Rev. B **31**(2), 805 (1985)
11. J. Bardeen. Tunnelling from a many-particle point of view. Phys. Rev. Lett. **6**(2), 57 (1961)
12. N.D. Lang. Spectroscopy of single atoms in the scanning tunneling microscope. Phys. Rev. B **34**(8), 5947–5950 (1986). doi:10.1103/PhysRevB.34.5947
13. A. Selloni, P. Carnevali, E. Tosatti, C.D. Chen. Voltage-dependent scanning-tunneling microscopy of a crystal surface: graphite. Phys. Rev. B **31**(4), 2602–2605 (1985). doi:10.1103/PhysRevB.31.2602
14. V.A. Ukraintsev. Data evaluation technique for electron-tunneling spectroscopy. Phys. Rev. B **53**(16), 11176–11185 (1996). doi:10.1103/PhysRevB.53.11176
15. J. Buisset, *Tieftemperatur Rastertunnelmikroskopie* (Wissenschaft and Technik Verlag, Berlin, 1996)
16. A.J. Heinrich. Single-atom spin-flip spectroscopy. Science **306**(5695), 466–469 (2004). doi:10.1126/science.1101077
17. Stanford Research Systems. Model SR810 DSP Lock-In amplifier—manual, January 2005. URL http://www.thinksrs.com/mult/SR810830m.htm
18. P. Wahl, L. Diekhoner, M.A. Schneider, K. Kern. Background removal in scanning tunneling spectroscopy of single atoms and molecules on metal surfaces. Rev. Sci. Instrum. **79**(4), 043104–043104-4 (2008). doi:10.1063/1.2907533
19. J.A. Stroscio. Imaging electronic surface states in real space on the Si(111) 2 × 1 surface. J. Vac. Sci. Technol., A: Vacuum, Surfaces, and Films **5**, 838 (1987). doi:10.1116/1.574321
20. J. Li, W. Schneider, R. Berndt. Local density of states from spectroscopic scanning-tunneling-microscope images: Ag(111). Phys. Rev. B **56**(12), 7656–7659 (1997). doi:10.1103/PhysRevB.56.7656
21. Y. Yayon, X. Lu, M.F. Crommie. Bimodal electronic structure of isolated Co atoms on Pt(111). Phys. Rev. B **73**(15), 155401 (2006). doi:10.1103/PhysRevB.73.155401
22. M. Ziegler, N. Néel, A. Sperl, J. Kröger, R. Berndt. Local density of states from constant-current tunneling spectra. Phys. Rev. B **80**(12), 125402 (2009). doi:10.1103/PhysRevB.80.125402
23. P. Avouris. Manipulation of matter at the atomic and molecular levels. Acc. Chem. Res. **28**(3), 95–102 (1995). doi:10.1021/ar00051a002
24. L. Gross, N. Moll, F. Mohn, A. Curioni, G. Meyer, F. Hanke, M. Persson. High-resolution molecular orbital imaging using a p-wave STM tip. Phys. Rev. Lett. **107**(8), 086101 (2011). doi:10.1103/PhysRevLett.107.086101
25. J.A. Stroscio, D.M. Eigler. Atomic and molecular manipulation with the scanning tunneling microscope. Science **254**(5036), 1319–1326 (1991). doi:10.1126/science.254.5036.1319
26. G. Meyer, K. Rieder. Controlled manipulation of single atoms and small molecules with the scanning tunneling microscope. Surf. Sci. **377–379**(0), 1087–1093 (1997). doi:10.1016/S0039-6028(96)01551--8
27. G. Meyer. A simple low-temperature ultrahigh-vacuum scanning tunneling microscope capable of atomic manipulation. Rev. Sci. Instrum. **67**(8), 2960 (1996). doi:10.1063/1.1147080

References

28. CreaTec Fischer and Co. GmbH. "STM/AFM systems",. http://www.createc.de/products/lt-stm-afm-systems.html. URL http://www.createc.de/products/lt-stm-afm-systems.html
29. G. Kresse and J. Furthmüller. Efficiency of ab-initio total energy calculations for metals and semiconductors using a plane-wave basis set. Comput. Mater. Sci. **6**(1), 15–50 (1996). doi:10.1016/0927-0256(96)00008--0
30. G. Kresse, D. Joubert, From ultrasoft pseudopotentials to the projector augmented-wave method. Phys. Rev. B **59**(3), 1758 (1999)
31. J.P. Perdew, A. Zunger. Self-interaction correction to density-functional approximations for many-electron systems. Phys. Rev. B 23(10), 5048 (1981). doi:10.1103/PhysRevB.23.5048
32. J.P. Perdew, K. Burke, M. Ernzerhof. Generalized gradient approximation made simple. Phys. Rev. Lett. **77**(18), 3865–3868 (1996). doi:10.1103/PhysRevLett.77.3865
33. S. Grimme. Semiempirical GGA-type density functional constructed with a long-range dispersion correction. J. Comput. Chem. **27**(15), 1787–1799 (2006). doi:10.1002/jcc.20495
34. T. Bučko, J. Hafner, S. Lebègue, J.G. Ángyán. Improved description of the structure of molecular and layered crystals: Ab initio DFT calculations with van der waals corrections. J. Phys. Chem. A **114**(43), 11814–11824 (2010). doi:10.1021/jp106469x
35. M. Dion, H. Rydberg, E. Schröder, D.C. Langreth, B.I. Lundqvist. Van der waals density functional for general geometries. Phys. Rev. Lett. **92**(24), 246401 (2004). doi:10.1103/PhysRevLett.92.246401
36. D.C. Langreth, B.I. Lundqvist, S.D. Chakarova-Käck, V.R. Cooper, M. Dion, P. Hyldgaard, A. Kelkkanen, J. Kleis, L. Kong, S. Li, P.G. Moses, E. Murray, A. Puzder, H. Rydberg, E. Schröder, T. Thonhauser. A density functional for sparse matter. J. Phys.: Condens. Matter **21**, 084203 (2009). doi:10.1088/0953-8984/21/8/084203
37. T. Thonhauser, V.R. Cooper, S. Li, A. Puzder, P. Hyldgaard, D.C. Langreth. Van der waals density functional: Self-consistent potential and the nature of the van der waals bond. Phys. Rev. B **76**(12), 125112 (2007). doi:10.1103/PhysRevB.76.125112
38. G. Román-Pérez, J.M. Soler. Efficient implementation of a van der waals density functional: application to double-wall carbon nanotubes. Phys. Rev. Lett. **103**(9), 096102 (2009). doi:10.1103/PhysRevLett.103.096102
39. S.L. Dudarev, G.A. Botton, S.Y. Savrasov, C.J. Humphreys, A.P. Sutton. Electron-energy-loss spectra and the structural stability of nickel oxide: an LSDA+U study. Phys. Rev. B **57**(3), 1505–1509 (1998). doi:10.1103/PhysRevB.57.1505
40. R.F.W. Bader, W.H. Henneker, P.E. Cade. Molecular charge distributions and chemical binding. J. Chem. Phys. **46**(9), 3341–3363 (1967). doi:10.1063/1.1841222
41. W. Tang, E. Sanville, G. Henkelman. A grid-based bader analysis algorithm without lattice bias. J. Phys.: Condens. Matter **21**(8), 084204 (2009). doi:10.1088/0953-8984/21/8/084204

Chapter 3
Introduction to the Kondo Effect

Conduction measurements of magnetic systems at low temperature, like the present study of 3d transition metal MePc on Ag(100) are closely intertwined with a phenomenon known as the Kondo effect. Essentially it is a coupling of the conduction electrons of a metal to a magnetic impurity. Below a critical temperature, the so-called Kondo temperature, the coupling leads to a screening of the spin of the magnetic impurity due to the creation of a many-body singlet state. This gives rise to a typical sharp resonance in the DOS at the Fermi level, which can be accessed experimentally by STS measurements. This Kondo resonance is characteristic of magnetic impurities on a non-magnetic surface. Hereby it provides a method for STM measurements to probe the magnetic properties of adatoms or molecules.

As the Kondo effect is intrinsically a many-body problem, its mathematical description is necessarily complex. Since its first explanation in the 1960, many advanced mathematical tools such as the Numerical Renormalization Group (NRG), and the Bethe ansatz have been applied to solve the problem. In this chapter we present an outline of the basic concepts in a descriptive manner, in order to understand the mechanisms involved. For a more complete treatment the reader is referred to the specific literature [1–4].

3.1 The Kondo Problem

In the 1930s measurements of the electrical resistivity of certain metals revealed an effect that would puzzle physicists for 30 years [6]: as the temperature of the sample is lowered, the resistivity reaches a minimum and then increases again as $-ln(T)$, for even lower temperature (see Fig. 3.1). Later it was found that this effect only occurs if the metal contains dilute magnetic impurities, such as iron or cobalt atoms.

In metals, the electric resistivity is caused by the scattering of conduction electrons. They can scatter off crystal lattice vibrations (phonons) or off static defects in the lattice itself such as vacancies or substitutions.

Fig. 3.1 Resistivity of a copper crystal, with different percentages of Fe impurities. Figure adapted from [5]

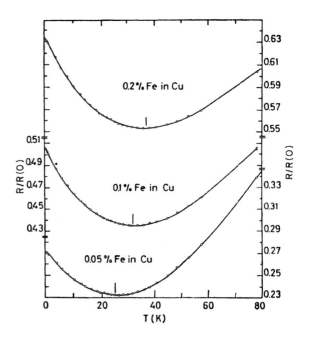

As the temperature decreases, lattice vibrations become less pronounced, hence the effect of phonon scattering on the resistivity goes down ($\propto T^5$). The contribution of the static impurities remains unchanged, and the resistivity saturates at a minimum caused by these defects.

The increase in resistivity observed for the dilute magnetic alloys was explained by J. Kondo in 1964 [7], by introducing another scattering mechanism for the conduction electrons. The spin flip interaction between the magnetic impurities and the spin of the conduction electrons. He described this in the s-d exchange model or Kondo model:

$$H = H_{Bloch} + H_K \tag{3.1}$$

$$H_K = -J\vec{S} \cdot \vec{s}(r) \tag{3.2}$$

The Hamiltonian consists of a H_{Bloch} part describing the non interacting electrons in a Bloch model, and the coupling part H_K which couples the spin $\vec{S}$ of the impurity to the spin $\vec{s}$ of the conduction electrons at the site of the impurity r. A Heisenberg type spin interaction is represented by J. Using this Hamiltonian in third order perturbation theory to determine the resistivity, J. Kondo was able to reproduce the $-ln(T)$ behavior observed in the dilute alloys. We will not go into details, because the calculations are rather lengthy and do not provide any further insight into the physics involved.

3.1 The Kondo Problem

There are however several important details to point out: Within the Kondo model J is assumed to be constant and positive, implying anti-ferromagnetic coupling between the impurity and the conduction electrons. This coupling sign is a necessary condition for the perturbation theory to produce the correct results. A more physical explanation can be found in the Schriefer-Wolff transformation [8] (see Sect. 3.3 on p. 36).

The Kondo model, however, has some limitations. The description of the impurity as a spin $1/2$ system, is with regard to real impurities an insufficient description. Higher spin or more complex mixed charge states for impurities are possible. A more problematic shortcoming is the breakdown of Kondo's model for very low temperatures. For $T \to 0$ it makes the unphysical prediction that the resistivity will be infinite. The Kondo perturbation theory description of the susceptibility breaks down at a finite temperature T_K. This critical temperature became known as the Kondo temperature [1]:

$$k_B T_K = D e^{-1/(2J\rho_o)} \tag{3.3}$$

D is the width and ρ_o the density of states of the conduction band. One has to be careful not to confuse the temperature of the resistivity minimum with T_K. When the system reaches T_K, the magnetic properties of the impurity start changing. As will be detailed later in Sect. 3.4, its moment is screened by the conduction electrons, forming a many-body singlet state.

The first solution to the Kondo problem was found by Anderson using a scaling approach, known as "poor man's scaling", which gives a qualitative understanding of the Kondo physics [9]. A theory that produced quantitative results is the "numerical renormalization group" method developed by Wilson in the mid seventies [10]. The discussion of these theories lies beyond the scope of this work, instead we will follow a more descriptive approach.

3.2 The Anderson Model

Kondo's spin flip scattering model for dilute alloys focuses on the phenomenological description of the resistivity minimum. A microscopic model describing a magnetic impurity inside a conducting metal is the Anderson Model [11]. It offers a very physical description, and thus helps to understand the underlying mechanism involved. The central question it tries to answer is whether a magnetic impurity interacting with a free-electron metal retains is magnetic character or not. The system is modeled by only three parameters, which makes it rather easy to handle as a numerical model. It was shown by Schrieffer and Wolf that it is possible to transform Anderson's into Kondo's Hamiltonian, and that for certain parameters the Anderson model describes the same physics as the Kondo model, as we will discuss later.

3.2.1 The Anderson Hamiltonian

The Anderson model reduces the problem of a dilute magnetic alloy to a single magnetic impurity placed in a non magnetic host metal (see Fig. 3.2). A magnetic atom would have unpaired electrons, e.g., in 3d or 4f states, which are represented as a single impurity state E_d, which can be either singly or doubly occupied. These states are quite localized, therefore the addition of an extra electron has to overcome a strong Coulomb repulsion. This behavior is modeled by introducing an additional potential U, reducing the probability of double occupation. As we will see later this term plays a very important role in the creation of a magnetic moment in the impurity. The formal Hamiltonian is then constructed as follows. The host metal is represented as non interacting Bloch electrons, the local impurity state is modeled as extra localized orbitals, idealized as one state with two possible spin orientations. Any degeneracy of this orbital is neglected. Another term describes the hybridization between the conduction band electrons and the impurity:

$$H_A = \underbrace{\sum_{\vec{k}\sigma} E_k n_{\vec{k}\sigma}}_{\text{conduction band}} + \underbrace{\frac{1}{\sqrt{N}} \sum_{\vec{k}\sigma} \left[V_{\vec{k}d} c^{\dagger}_{\vec{k}\sigma} c_{d\sigma} + V^*_{\vec{k}d} c^{\dagger}_{d\sigma} c_{\vec{k}\sigma} \right]}_{\text{hybridization}} \quad (3.4)$$

$$+ \underbrace{E_d(n_{d\uparrow} + n_{d\downarrow})}_{\text{impurity state}} + \underbrace{U n_{d\uparrow} n_{d\downarrow}}_{\text{coulomb repulsion}}$$

where E_k is the energy level of the conduction band state, $n_{\vec{k}\sigma} = c^{\dagger}_{\vec{k}\sigma} c_{\vec{k}\sigma}$ is the electron number operator, while $c^{\dagger}_{\vec{k}\sigma}, c_{\vec{k}\sigma}$ refer to the respective creation and annihilation operators, σ is respective the spin state. E_d is the impurity state energy with $n_{d\uparrow}$ and

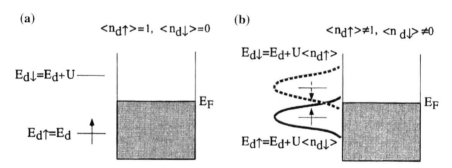

Fig. 3.2 A schematic view of the Anderson model. The impurity is modeled as a singly occupied localized state, double occupation is inhibited by an on-site Coulomb repulsion U. The metal is a continuum of states filled up to E_F. **a** The hybridization of the local levels is neglected, while the Coulomb repulsion is already considered. **b** The complete Anderson model: the hybridization is switched on leading to a broadening of the impurity states, which to some extent equalizes the occupancy of spin up and spin down level. Image adapted from [3]

3.2 The Anderson Model

$n_{d\downarrow}$ again the number operators for the impurity orbital (with $c_{d\sigma}^{\dagger}$, $c_{d\sigma}$ creation and annihilation operators). The hybridization term depends on the transition amplitude $V_{\vec{k}\sigma}$ and is normalized by the number units cells of the host metal N. The strength of the Coulomb repulsion is denoted as $U > 0$.

The behavior of the impurity is thus determined by three parameters:

- the **on-site Coulomb repulsion** U at the impurity, which favors a singly occupied impurity orbital and thereby a persisting moment.
- the **position of the impurity energy level** E_d with respect to the Fermi energy E_F of the host metal. If $E_d < E_F$ and $E_d + U > E_F$ the impurity level will be singly occupied, while $E_d < E_d + U < E_F$ would lead to a doubly occupied impurity level.
- the **hybridization** between the local and conduction states causing a quenching of the moment by smearing out the occupation of the impurity level, as we shall see in the next section.

While the effects of the two first parameters are rather straightforward to identify, the hybridization requires more consideration. From an intuitive point of view the consequence of hybridization between a localized state E_d and the continuum leads to a broadening of the former impurity levels. Indeed the treatment of the Anderson Hamiltonian leads to a Lorentzian line shape for the density of states of the impurity levels, with the broadening depending on the hybridization term $V_{\vec{k}\sigma}$.

Using a Green's function $G^0(E) = (E - H_A)^{-1}$ approach and a Hartee- Fock approximation to simplify the operator $U n_{d\uparrow} n_{d\downarrow} \approx U n_{d\uparrow} \langle n_{d\downarrow} \rangle + U n_{d\downarrow} \langle n_{d\uparrow} \rangle$ one finds the occupation of the impurity levels to be [11]:

$$n_{d\uparrow(\downarrow)}(E) = \frac{1}{\pi} \frac{\Gamma}{\left(E - E_d - U \langle n_{d\downarrow(\uparrow)} \rangle\right)^2 + \Gamma^2} \quad (3.5)$$

with $\Gamma = \pi \left|V_{\vec{k}\sigma}\right|^2 n_o(E_F)$

The term $n_o(E_F)$ denotes the electronic density of states of the host at the Fermi-level for a given spin direction per unit cell. Γ will be the parameter characterizing the hybridization.

Using these terms to integrate the density of states up to the Fermi-level, the occupation of the impurity level can be found:

$$\langle n_{d\downarrow} \rangle = \int_{-\infty}^{E_F} n_{d\uparrow}(E) dE \quad (3.6)$$

$$\langle n_{d\uparrow} \rangle = \int_{-\infty}^{E_F} n_{d\downarrow}(E) dE \quad (3.7)$$

The solutions have to be determined graphically, because the resulting equations are still coupled. Note that, if $\Gamma \gg U$ only non-magnetic solutions $\langle n_{d\downarrow} \rangle = \langle n_{d\uparrow} \rangle = 1/2$ appear. Equally if $\Gamma \ll U$ only the magnetic solution is found

Fig. 3.3 Phase diagram showing all the magnetic and non-magnetic states as predicted by the Hartree Fock treatment of the Anderson model. Image adapted from [11]

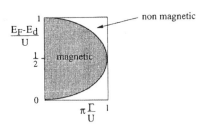

$\langle n_{d\uparrow}\rangle \approx 1, \langle n_{d\downarrow}\rangle \approx 0$. The transition between the magnetic and non-magnetic solution occurs at $\pi\Gamma = U$. Finally one obtains a phase diagram showing all magnetic and non magnetic solutions for different sets of parameters for the Anderson Hamiltonian, shown in Fig. 3.3.

In conclusion the Anderson model describes the stability of a magnetic impurity in a non magnetic host metal. It includes two types of counteracting potentials: first the on site Coulomb repulsion which leads to singly occupied impurity state, which maintains the magnetic moment. Counteracting this effect is the hybridization between the impurity state and the conduction electrons, smearing out the occupation and effectively reducing the magnetic moment.

It is important to note that all these considerations are only valid at high enough temperatures $T > T_K$. As the thermal energy of the systems is lowered, any remaining magnetic moment will be screened by strong electron correlation effects, i.e., the Kondo effect (see Sect. 3.4 on p. 38).

3.3 Virtual Spin Flips and the Schrieffer: Wolff Transformation

As described in the last section the Anderson model predicts that the local moment of a magnetic impurity persists if the on-site Coulomb repulsion is larger than the hybridization energy. In this case it essentially describes the same physics as the Kondo model. Schrieffer and Wolff have shown that it is possible to obtain the Kondo Hamiltonian from the Anderson model, in its magnetic limit, by diagonalizing it in the subspace of singly occupied impurity states [8] . This is called the Schrieffer Wolff transformation.

Before entering into details, we would like to present an intuitive way of picturing the Kondo model in terms of the Anderson model, i.e., the Schrieffer Wolff transformation. Consider the following situation in the Anderson model: the impurity level E_d lies well below E_F and the doubly occupied counterpart above the Fermi level ($E_d + U > E_F$). The temperature is sufficiently low to exclude thermal excitations, $k_B T \ll |E_d|, E_d + U$. This situation leads to a singly occupied impurity level. As the energy of the system is low, no classical exchange of the impurity and the conduction spin is possible. However, quantum mechanics permits two virtual excitations that can flip the spin of the impurity: In Fig. 3.4a one electron tunnels from the magnetic

3.3 Virtual Spin Flips and the Schrieffer: Wolff Transformation

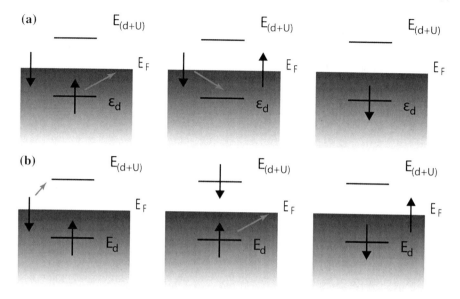

Fig. 3.4 *Two* processes that contribute to the second order fluctuations on the impurity, flipping its spin. **a** The conduction electron tunnels through the hybridized level into the impurity, creating a double occupied intermediate state. Afterwards the original impurity electron leaves the impurity. **b** The impurity electron leaves and is then replaced by a conduction electron. U is the Coulomb repulsion, d is the impurity level. Image adapted from [4]

atom to the conduction band in a state near the Fermi level, which are the only ones available due to the low temperature. Another electron tunnels back into the impurity. The new electron can now have its spin oriented the other way, thus the impurity spin is effectively flipped without any energy cost. Applying the same arguments another spin flip process is possible, shown in Fig. 3.4b. An electron from the conduction band tunnels into the impurity, temporarily doubly occupying it. According to the Pauli principle the spin of the incoming electron has to be opposite to the one of the impurity electron. If now the original localized electron tunnels out, the impurity spin is flipped again.

This intuitive reasoning outlines how, at low temperature, the Anderson model can be transformed into a model that considers only one effective spin in the impurity, which can be flipped by virtual processes with the conduction electrons. Hence, in the case of a singly occupied level, the impurity can be described by its spin $\vec{S}$. The coupling to the spin $\vec{s}$ of the conduction electrons is proportional to a coupling constant J. The interaction is thus found to be $J\vec{S}\vec{s}$, which is exactly the Kondo model.

The actual mathematical Schrieffer-Wolff transformation of the Anderson to Kondo Hamiltonian is a canonical transformation in lowest order [4]. It predicts the sign of the coupling constant J to be anti-ferromagnetic, which in Kondo's model was a parameter. The mathematical form of the transformation S can be written as [8]:

$$S = \sum_{k,\sigma,\alpha=\pm} \frac{V_{k,d}}{E_k - E_\alpha} n_{d,-\sigma}^\alpha c_{k,\sigma}^\dagger d_\sigma - h.c.$$

with : $E_+ = E_d + U$; $E_- = E_d$

$$n_{d,-\sigma}^- = 1 - n_{d,-\sigma}; \quad n_{d,-\sigma}^+ = n_{d,-\sigma}$$

The Anderson Hamiltonian H_A is then transformed with:

$$H_A = H_0 + H_1$$
$$H_{SW} = H_0 + \frac{1}{2}[S, H_1]$$

where H_{SW} denotes the transformed Hamiltonian, H_1 the perturbed part of the Anderson Hamiltonian, which is the hybridization of the local level and the conduction band. H_0 is the rest of the Anderson model, i.e., the conduction band, the Coulomb repulsion and the localized level.

Through this approach it can be shown that the coupling constant J is given by [4]:

$$J \approx \frac{U}{|E_d|(U - |E_d|)} < 0 \tag{3.8}$$

Thus charge fluctuations, i.e., virtual excitations in the Anderson model at low temperature lead to a spin dependence of the interaction. The Pauli principle is the reason for this, as it forbids to doubly occupy the impurity with electrons of the same spin. We have seen, how the transformed Anderson model leads to a Kondo model with an anti-ferromagnetic exchange coupling, resulting in the condensation of the system in a singlet state, as will be discussed in the next section.

3.4 Formation of the Kondo Singlet

Experiments on dilute magnetic alloy revealed yet another "anomalous" behavior in systems with a resistivity minimum. The magnetic susceptibility ceases to follow Curie's law and saturates to a constant value for $T = 0$, indicating that the magnetic moment is completely quenched at low temperature. This behavior can be explained by the formation of a many-body singlet state formed by the impurity and the surrounding conduction electrons at the Fermi energy.

3.4.1 The Spin 1/2 Kondo Effect

A Molecular Model

The screening of the local moment is already included in the Anderson model. The mechanism involved can be nicely understood in a simplified description as a two

3.4 Formation of the Kondo Singlet

orbital molecule [3]. We make an extreme simplification by treating the host metal as just one *extended* (ϵ_k) orbital and the impurity as a single localized orbital (ϵ_f). We use an Anderson type Hamiltonian, and consider two electrons inside the molecule. The parameter of the model are: the Coulomb repulsion U for the localized orbital and the hybridization parameter V_{sf}. In the case of zero Coulomb repulsion ($U = 0$) this model describes the two orbitals with the energies ϵ_k and ϵ_f interacting with the off diagonal element V_{sf}.

$$\begin{vmatrix} \epsilon_k - E & V_{sf} \\ V_{sf} & \epsilon_f - E \end{vmatrix} = 0 \qquad (3.9)$$

The diagonalization of this Anderson Hamiltonian leads to bonding and antibonding states with the energies E_a and E_b. The hybridization between the two orbitals is assumed to be small $|V_{sf}| \ll \epsilon_k - \epsilon_f = \Delta$ thus the energies will be respectively:

$$E_b = \epsilon_f - \frac{|V_{sf}|^2}{\Delta} \qquad E_a = \epsilon_k + \frac{|V_{sf}|^2}{\Delta} \qquad (3.10)$$

The two electrons can now be distributed in six different ways into these two orbitals, creating one triplet (S = 1) and three singlet (S = 0) states, as shown in Fig. 3.5. The ground state will be the configuration with 2 electrons in the bonding orbital forming a singlet state.

If the on-site repulsion is switched on, and for the sake of simplicity we use $U = \infty$, the energy landscape changes significantly. There are five possible distributions for the electrons in the two orbitals, because the double occupation of the localized orbital is forbidden (see Fig. 3.6).

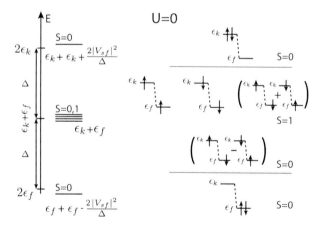

Fig. 3.5 The energy levels of the molecular model, for Coulomb repulsion being $U = 0$. ϵ_k stands for the extended orbital (a crude model of the conduction band), while ϵ_f for the localized orbital (i.e. the impurity), hybridization is assumed to be small $|V_{sf}| \ll \epsilon_k - \epsilon_f = \Delta$. The possible two-electron occupation configurations are shown on the right creating one triplet and three singlet states

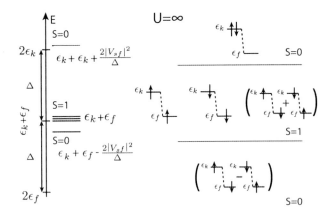

Fig. 3.6 The energy levels of the molecular model, for a Coulomb repulsion $U = \infty$. The possible two electron occupation configurations are shown on the *right*. The double occupation for the impurity ϵ_f is forbidden by $U = \infty$, creating one triplet and two singlet states

Two singlet states can be formed:

$$\psi_1^{(S=0)} = c_{k\uparrow}^\dagger c_{k\downarrow}^\dagger |0\rangle, \qquad \psi_2^{(S=0)} = \frac{1}{\sqrt{2}}\left[c_{k\uparrow}^\dagger c_{f\downarrow}^\dagger - c_{k\downarrow}^\dagger c_{f\uparrow}^\dagger\right]|0\rangle \qquad (3.11)$$

with the corresponding Hamiltonian:

$$H_m = \begin{pmatrix} 2\epsilon_k & \sqrt{2}V_{sf} \\ \sqrt{2}V_{sf}^* & \epsilon_k + \epsilon_f \end{pmatrix} \qquad (3.12)$$

giving the eigenvalues:

$$E = \epsilon_k + \frac{1}{2}\left[\epsilon_k + \epsilon_f \pm \sqrt{(\epsilon_k - \epsilon_f)^2 + 8|V_{sf}|^2}\right] \qquad (3.13)$$

which then by a series expansion and the assumption $|V_{sf}| \ll \epsilon_k - \epsilon_f = \Delta$ can be written as:

$$E^{(S=0)} = \epsilon_k + \epsilon_f - \frac{2|V_{sf}|^2}{\Delta} \qquad E^{(S=0)} = \epsilon_k + \epsilon_k + \frac{2|V_{sf}|^2}{\Delta} \qquad (3.14)$$

One triplet can also be formed, with the wavefunctions for $S_z = 1, 0, -1$:

$$c_{k\uparrow}^\dagger c_{f\uparrow}^\dagger |0\rangle, \qquad \frac{1}{\sqrt{2}}\left[c_{k\uparrow}^\dagger c_{f\downarrow}^\dagger + c_{k\downarrow}^\dagger c_{f\uparrow}^\dagger\right]|0\rangle, \qquad c_{k\downarrow}^\dagger c_{f\downarrow}^\dagger |0\rangle$$

The energy of the triplet state is accordingly:

3.4 Formation of the Kondo Singlet

$$E^{(S=0)} = \epsilon_k + \epsilon_f \tag{3.15}$$

By combining Eqs. 3.14 and 3.15 we find the energy levels (see Fig. 3.6):

$$E^{(S=0)} = \epsilon_k + \epsilon_f - \frac{2|V_{sf}|^2}{\Delta} < E^{(S=1)} = \epsilon_k + \epsilon_f$$
$$< E^{(S=0)} = \epsilon_k + \epsilon_k + \frac{2|V_{sf}|^2}{\Delta} \tag{3.16}$$

It becomes clear that the ground state of the system is a singlet state, which lies lower in energy than the triplet. The spin of the localized level is effectively screened. The energy difference between them $\frac{2|V_{sf}|^2}{\Delta}$ is, however, rather small, it corresponds to one spin flip to form the triplet state. If the thermal energy of the system is high enough, i.e., larger than $k_B T_K = \frac{2|V_{sf}|^2}{\Delta}$ these flips can be pictured as spin fluctuations. Hence as the temperature is lowered the spin fluctuations between the ground singlet and the triplet state cease to occur and the singlet state becomes predominant.

We have found this result using a dramatically simplified description of the conduction band: the mechanism for the formation of the singlet ground state is in essence the same when a complete band is considered. There is one very important difference, however: The true Kondo ground state is not formed by the impurity electron and a single conduction electron. The screening of the impurity moment is a many-body effect, involving all conduction electrons close to the Fermi energy.

The Kondo Temperature

In light of this model we can now better understand the Kondo temperature, which was defined earlier as the breakdown of the Kondo model (Eq. 3.3). The molecular model showed that below a certain temperature the magnetic moment is screened because the system can lower its energy by forming a singlet state. This state breaks up if the thermal energy of the system is higher than the energy gained by its formation. The Kondo temperature is defined as the energy gained by the system through the formation of the Kondo singlet. Exact treatment of the Anderson model with an advanced mathematical method like the NRG finds the following expression [12]:

$$T_K = \frac{1}{2}(\Gamma U)^2 exp\left(\frac{\pi E_d(E_d+U)}{\Gamma U}\right) \tag{3.17}$$

where Γ is the width of the impurity energy level E_d and U the on-site Coulomb repulsion.

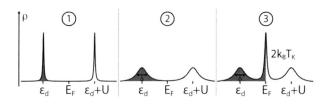

Fig. 3.7 Density of states in the Anderson model: (1) the unperturbed impurity levels. (2) The impurity interacting with the conduction sea well above the Kondo temperature. (3) and well below the Kondo temperature. The many-body singlet causes the sharp peak at the Fermi energy

The Zero Bias Kondo Resonance

The formation of the singlet state has another consequence for the system, as first noticed by Abrikosov [13] and Suhl [14]: a very sharp resonance appears in the DOS exactly at the Fermi level, its half width at $T = 0$ is $k_B T_K$ (Fig. 3.7).

In the complete Anderson model a tangible qualitative picture can be created: The host electrons, with an energy $E = k_B T_K$ around E_F constantly tunnel in and out of the impurity state along the lines of the argumentation for the Schrieffer Wolff transformation, creating a new quasi particle state. This can be thought of as a cloud of electrons, whose spin is correlated due to the interaction with the impurity spin and the Pauli principle. Since this spin flip mechanism does not require any energy, the resonance is pinned at the Fermi level. Its width Γ_K depends on the Kondo temperature as well as the measurement temperature, and follows an expression derived from Fermi liquid theory [15, 16]:

$$\Gamma_K(T) = 2\sqrt{(\pi k_B T)^2 + 2(k_B T_K)^2}. \tag{3.18}$$

3.4.2 The Underscreened Kondo Effect

So far we have considered impurities with a total spin of $S = 1/2$, which at temperatures far below T_K are completely screened by the conduction electrons. What happens, however, when the total spin is greater than $1/2$, e.g., for a $S = 1$ system?

The behavior of such a system depends on the number of screening channels and on its temperature. The overlap of the wave functions of the impurity and the conduction electrons leads to the creation of screening channels. If the number n of channels is high enough ($n = 2S$), even large magnetic moments will be completely screened [17].

Consider an $S = 1$ impurity with two screening channels, which are different due to different overlap with the conduction electrons. Both channels will have their own antiferromagnetic Kondo coupling constant $J_{1,2}$, which depend on the hybridization Γ with the conduction electrons. Consequently Eq. 3.3 will give two Kondo

3.4 Formation of the Kondo Singlet

temperatures T_{K_1} and T_{K_2}. The coupling is asymmetric, so that one Kondo temperature will be larger than the other $T_{K_1} < T_{K_2}$. The behavior of such a system, then depends on the temperature. If $T \ll T_{K_1} < T_{K_2}$ the spin will be completely screened, a situation known as a two-stage Kondo effect. An underscreened scenario occurs for $T_{K_1} < T < T_{K_2}$. The spin of the impurity will be only partially screened from S = 1 to S = 1/2, because only one channel is "active".

This situation has an impact on the intensity G_K of the Kondo resonance as a function of temperature. For the fully screened Kondo effect, the conductance increases logarithmically above the Kondo temperature and below it, saturates in a quadratic fashion. On the other hand for the underscreened Kondo effect two different logarithmic behaviors above and below the higher Kondo temperature T_{K_2} occur [18–20]. This is best seen by plotting the derivative of the conductance with respect to temperature as a function of inverse temperature. In Fig. 3.8 the behavior of the intensity of the Kondo resonance for a normal and an underscreened process is depicted.

The temperature dependent conductance obtained by numerical renormalization group (NRG) can be described using the following empirical formula [20, 21]:

$$G_K(T) = G_{\text{off}} + G(0) \cdot \left[1 + \left(\frac{T}{T_K} \right)^\xi \cdot (2^{1/\alpha} - 1) \right]^{-\alpha} \quad (3.19)$$

where Γ_K is the width (FWHM) and G_K the intensity of the Kondo resonance, α and ξ are parameters that depend on the total spin of the molecule (see Table 3.1).

Examples that show this kind of behavior are a C_{60} molecule placed off centered between two metallic leads, with a different coupling constant for each of them [20], Co(tpy-SH)2 complexes in break-junction experiments [21], or molecules adsorbed

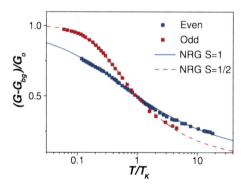

Fig. 3.8 Intensity of the Kondo resonance versus the inverse temperature recorded for a C_{60} molecule coupled to two metallics leads. The electron occupation of the molecule can be changed from odd to even leading to different spin states, with a fully screened and an underscreened Kondo effect. The curves are fitted with Eq. 3.19. Adapted from [20]

Table 3.1 Fitted parameters determined for 3.19 to approximate the NRG predictions for the temperature dependence of the Kondo resonance for an impurity with spin S and a single screening channel

Spin	ξ	α
S = 1/2	2	0.220 ± 0.005
S = 1	0.745 ± 0.009	0.506 ± 0.009

Adapted from [21]

on an surface, where two magnetic moments are differently coupled to the surface electrons (this work).

3.4.3 The Inelastic Kondo Effect

So far we have discussed the equilibrium Kondo effect, which leads to a sharp resonance at zero-bias. It is however possible for the Kondo interaction to couple to inelastic excitations. The coupling to vibrational [22–25], magnetic excitations [21, 26–29], photon adsorption/emission [30], cooper pair formation [31] has been observed. In molecules only the coupling to vibrational, and magnetic excitation persists, due to the large electronic level spacing $\Delta E \gg T_K$, and the absence of light.

Depending on the molecular orbital, the electrons can couple to the vibrations of the molecule while tunneling through it. For the Kondo effect, the basic principle can be understood in the virtual spin flip picture described on p. 36. An excited electron or hole tunnels into the impurity, but at the same time excites a phonon. These processes will create two additional sidebands above and below E_F at the energy of the excited phonon [23]. One has to bear in mind that these kind of features will be superposed to the normal inelastic coupling to phonons, which gives rise to step like conduction features, and of course the actual impurity level (see Fig. 3.9 for an example).

Magnetic excitations occur in much the same way. For instance in a system with a singlet ground state, the tunneling electrons can induce spin-flip tunneling when the applied bias is large enough to induce transitions to triplet states. Such processes give rise to Kondo resonance peaks out of equilibrium [28, 29].

3.5 The Kondo Effect in STM Measurements

In recent years scientific interest in the Kondo effect has been renewed. Thanks to new experimental techniques, it became possible to study Kondo systems at the nanoscale, for instance in single atoms or molecules. Break-junction, quantum dot and STM measurements allow an unprecedented control over physical parameters, and have led to the discovery of many new forms of Kondo behavior, inaccessible in bulk studies of diluted alloys. An extensive review of these devices is given in Refs. [32, 33], here we will present a few examples of STM studies.

3.5 The Kondo Effect in STM Measurements

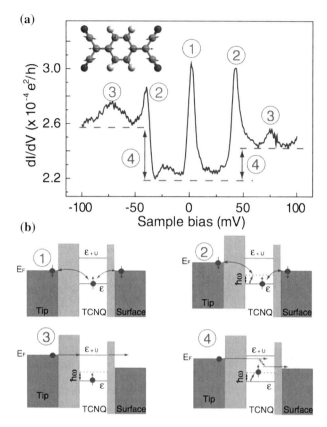

Fig. 3.9 **a** dI/dV spectrum taken on the center of a tetracyanoquinodimethane TCNQ molecule adsorbed on Au(111). Various elastic and inelastic processes can be identified. The inset shows the motion of TCNQ carbon atoms for the 9 vibrational modes. The arrows are proportional to the amplitude of motion. **b** Schematics of the Kondo (1) and the phonon-assisted inelastic Kondo (2) tunneling, observed as sharp peaks in the spectrum (**a**). Energy diagram of resonant tunneling through the doubly occupied LUMO state (3) and inelastic (off-resonance) tunneling (4), observed as broad peaks and steps in the spectrum (**a**), respectively. Similar diagrams can be constructed for negative sample bias. Image adapted from [22]

STM measurements of Kondo system allow access to real space images of Kondo resonances, providing the possibility to investigate the effect directly for a single atom, while at the same time being able to manipulate the structure on the surface, investigating the relationship between geometrical arrangements, chemical bonding, and the Kondo interaction. Measurements of single magnetic adatoms on noble metal surfaces served as a proof of principle that STS was able to detect a Kondo resonance

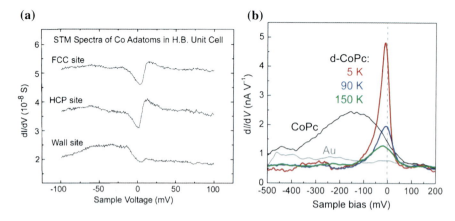

Fig. 3.10 **a** dI/dV spectra acquired for a single Co adatom positioned on different adsorption sites on the Au(111) herringbone unit cell. The Kondo resonance remains the same for fcc and hcp sites, but the Fano line shape undergoes a sharp change when the adatom is positioned on a dislocation wall. Image adopted from [37]. **b** dI/dV spectra taken at the center of a CoPc molecule at 5 K, adsorbed on Au(111) showing the appearance of a Kondo resonance for deprotonized d-CoPc molecule at 5, 90, and 150 K (*colored lines*). Spectra from bare Au(111) (*gray line*) is shown for comparison. Image adopted from [38]

at E_F [34, 35]. It became apparent that the line shape (see Fig. 3.10a), was not always a peak, but rather a spectral feature of more complex shape as we will discuss later[1].

The local environment of the impurity has a very strong impact on the shape of the Kondo resonance and temperature. For example, the change from 3 to 4 neighboring atoms, for a Co atom adsorbed on Cu(111) or Cu(100) surface changes T_K by almost 40 % [36].

The influence of a change in coupling between the magnetic atom and the substrate metal has been further explored by modifying its chemical environment. One approach was to place the magnetic ion inside an organic molecule. Zhao and coworkers found that Co phthalocyanine molecules on Au(111) are coupled in such a way that no Kondo effect can occur, however when the ligand of the molecule is deprotonized the effective Co - substrate distance becomes larger and the Kondo resonance appears, as shown in Fig. 3.10b [38]. Also within the molecular configuration the local density of states of the substrate plays an important role, e.g. for Fe phthalocyanine molecules on Au(111) an adsorption site specific Fano line shape and Kondo temperature have been observed [39].

In a more systematic approach the coupling of the impurity spin to the substrates conduction electrons was increased by bonding a different number of CO ligands to a Co atom and hereby increasing T_K [40]. A recent study on CoCu$_N$ clusters showed that the addition of adatoms affects the system in a more complex way than a simple

---

[1] Note that not all features at E_F are related to the Kondo effect. However the characteristic increase in width of the resonance peak with temperature, as well as its intensity evolution can be used as proof that a Kondo system is measured [15, 16], see also Eqs. 3.18 and 3.19.

3.5 The Kondo Effect in STM Measurements

decoupling and that the local and anisotropic electronic structure has to be taken into account [41].

The purposeful manipulation of the density of states at the Fermi level has also been shown to have an influence on T_K. The Kondo resonance of Mn-phthalocyanine molecules on Pb islands grown on Si(111) manifested oscillating Kondo temperatures as a function of the Pb film thickness. The crossing of quantum-well states led to oscillations in the density of states at E_F [42]. A similar result has been obtained for Co atoms on Cu/Co/Cu(100) multilayers [43].

Using the manipulation capabilities of an STM setup the exchange interaction of two magnetic atoms in relation to their distance can be investigated using the Kondo effect as a local probe. For Co atoms on Cu(100) [44], the coupling was found to change with the distance between the Co atoms from ferromagnetic to antiferromagnetic for increasing distances.

The Fano Line Shape

In Sect. 2.2 on p. 16 we have seen that using an STM it is possible to record a signal that is proportional to the DOS of the sample. In theory the sharp Kondo feature at E_F should also be accessible by STS if the system is measured at temperatures below T_K. Indeed in many STS experiments of Kondo systems a strong feature at E_F has been observed. However, the line shape of this feature varies considerably and is not generally a peak, but rather a Fano function. This function was first used to describe the interference of a discrete state with a continuum [45].

The exact expression of the Fano function is given by [45]:

$$\frac{dI}{dV}(\omega) = a \cdot \frac{(q + \epsilon(\omega))^2}{1 + \epsilon(\omega)^2} + b + c \cdot \omega \quad \text{with } \epsilon = \frac{\omega - \epsilon_k}{\Gamma} \quad (3.20)$$

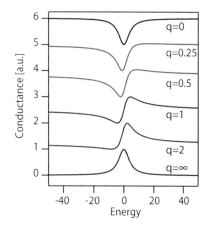

Fig. 3.11 Set of curves calculated with the Fano Eq. 3.20 for different values of q. At q = 0 a Lorentzian dip is detected which corresponds to mainly direct tunneling. In the limit of a large q, a Lorentzian peak is formed corresponding to mainly indirect tunneling. Intermediate values of q result in an S-like *curve*. All curves are shifted and normalized for clarity

where a, b, c are fitting constants, and ω is the energy. The factor q determines the line shape, which can be a dip for $q = 0$, an asymmetric feature ($q = 1$) or a peak ($q > 2$). For $q = \infty$ Eq. 3.20 becomes a Lorentzian function (see Fig. 3.11). Γ is the half width of the peak which is related to the Kondo temperature by $\Gamma \approx k_B T_K$. We can therefore use this function to fit the experimental dI/dV peaks, to find T_K. One has to bear in mind the additional broadening introduced by the temperature of the system (see Eq. 3.18).

For the interpretation of the q values found in experiments several models have been proposed. The first model is based on the coherent superposition of two different tunneling channels: a direct coupling channel to the impurity level, and an indirect channel probing the Kondo resonance via the substrate states. The picture implies that the line shape depends on the position of the tip as the contribution of the different tunneling channels would change. A different view states that in STS measurements the tip interacts much more with the substrate than with the very localized adsorbate states. Conductance measurements probe the presence of a magnetic atom through modified electronic properties of the metal surface. The Fano line shape is therefore thought of as arising from the interference of the Kondo resonance with the continuum of the metal states [46, 47].

References

1. A.C. Hewson, *The Kondo Problem to Heavy Fermions* (Cambridge University Press, Cambridge, 1997). ISBN 9780521599474
2. P. Fulde, *Electron Correlations in Molecules and Solids* (Springer, Berlin, 1995). ISBN 9783540593645
3. G. Grosso, G.P. Parravicini, *Solid State Physics*, 1st edn. (Academic Press, London, 2000). ISBN 012304460X
4. P. Phillips, P. Phillips, *Advanced Solid State Physics*, 1st edn. (Westview Press, Boulder, 2002)
5. J.P. Franck, F.D. Manchester, D.L. Martin, The specific heat of pure copper and of some dilute Copper+Iron alloys showing a minimum in the electrical resistance at low temperatures. Proc. R. Soc. Lond. A **263**(1315), 494–507 (1961). doi:10.1098/rspa.1961.0176
6. W. de Haas, J. de Boer, G. van dën Berg, The electrical resistance of gold, copper and lead at low temperatures. Physica **1**(7–12), 1115–1124 (1934). doi:16/S0031-8914(34)80310--2
7. J. Kondo, Resistance minimum in dilute magnetic alloys. Progress Theoret. Phys. **32**(1), 37–49 (1964)
8. J.R. Schrieffer, P.A. Wolff, Relation between the anderson and Kondo hamiltonians. Phys. Rev. **149**(2), 491 (1966). doi:10.1103/PhysRev.149.491
9. P.W. Anderson, A poor man's derivation of scaling laws for the Kondo problem. J. Phys. C: Solid State Phys. **3**(12), 2436–2441 (1970). doi:10.1088/0022-3719/3/12/008
10. K.G. Wilson, The renormalization group: Critical phenomena and the Kondo problem. Rev. Mod. Phys. **47**(4), 773 (1975). doi:10.1103/RevModPhys.47.773
11. P.W. Anderson, Localized magnetic states in metals. Phys. Rev. **124**(1), 41–53 (1961)
12. F.D.M. Haldane, Theory of the atomic limit of the anderson model. i. perturbation expansions re-examined. J. Phys. C: Solid State Phys. **11**(24), 5015–5034 (1978). doi:10.1088/0022-3719/11/24/030
13. A. Abrikosov, Magnetic impurities in metals: the sd exchange model. Physics **2**, 5 (1965)
14. H. Suhl, Dispersion theory of the Kondo effect. Phys. Rev. **138**, A515–A523 (1965). doi:10.1103/PhysRev.138.A515

15. K. Nagaoka, T. Jamneala, M. Grobis, M.F. Crommie, Temperature dependence of a single Kondo impurity. Phys. Rev. Lett. **88**(7), 077205 (2002). doi:10.1103/PhysRevLett.88.077205
16. M. Ternes, A.J. Heinrich, W.D. Schneider, Spectroscopic manifestations of the Kondo effect on single adatoms. J. Phys.: Condens. Matter **21**, 053001 (2009)
17. T.A. Costi, L. Bergqvist, A. Weichselbaum, J. von Delft, T. Micklitz, A. Rosch, P. Mavropoulos, P.H. Dederichs, F. Mallet, L. Saminadayar, C. Bäuerle, Kondo decoherence: Finding the right spin model for iron impurities in gold and silver. Phys. Rev. Lett. **102**(5), 056802 (2009). doi:10.1103/PhysRevLett.102.056802
18. P. Nozieres, A. Blandin, Kondo effect in real metals. J. Phys. **41**, 193–211 (1980)
19. A. Posazhennikova, P. Coleman, Anomalous conductance of a spin-1 quantum dot. Phys. Rev. Lett. **94**(3), 036802 (2005). doi:10.1103/PhysRevLett.94.036802
20. N. Roch, S. Florens, T.A. Costi, W. Wernsdorfer, F. Balestro, Observation of the underscreened Kondo effect in a molecular transistor. Phys. Rev. Lett. **103**(19), 197202 (2009). doi:10.1103/PhysRevLett.103.197202
21. J.J. Parks, A.R. Champagne, T.A. Costi, W.W. Shum, A.N. Pasupathy, E. Neuscamman, S. Flores-Torres, P.S. Cornaglia, A.A. Aligia, C.A. Balseiro, G.K. Chan, H.D. Abruña, D.C. Ralph, Mechanical control of spin states in spin-1 molecules and the underscreened Kondo effect. Science **328**(5984), 1370–1373 (2010). doi:10.1126/science.1186874
22. I. Fernández-Torrente, K.J. Franke, J.I. Pascual, Vibrational Kondo effect in pure organic Charge-Transfer assemblies. Phys. Rev. Lett. **101**(21), 217203 (2008). doi:10.1103/PhysRevLett.101.217203
23. J. Paaske, K. Flensberg, Vibrational sidebands and the Kondo effect in molecular transistors. Phys. Rev. Lett. **94**(17), 176801 (2005). doi:10.1103/PhysRevLett.94.176801
24. L. Yu, Z. Keane, J. Ciszek, L. Cheng, M. Stewart, J. Tour, D. Natelson, Inelastic electron tunneling via molecular vibrations in Single-Molecule transistors. Phys. Rev. Lett. **93**, 266802–266802 (2004)
25. T. Choi, S. Bedwani, A. Rochefort, C. Chen, A.J. Epstein, J.A. Gupta, A single molecule Kondo switch: multistability of tetracyanoethylene on Cu(111). Nano Lett. **10**(10), 4175–4180 (2010)
26. A. Kogan, G. Granger, M.A. Kastner, D. Goldhaber-Gordon, H. Shtrikman, Singlet-triplet transition in a single-electron transistor at zero magnetic eld. Phys. Rev. B **67**, 113309–113309 (2003)
27. E.A. Osorio, K. Moth-Poulsen, H.S.J.v.d. Zant, J. Paaske, P. Hedegaard, K. Flensberg, J. Bendix, T. Bjornholm, Electrical manipulation of spin states in a single electrostatically gated Transition-Metal complex. Nano Lett. **10**(1), 105–110 (2009)
28. J. Paaske, A. Rosch, P. Wölfle, N. Mason, C.M. Marcus, J. Nygård, Non-equilibrium singlet-triplet Kondo effect in carbon nanotubes. Nat. Phys. **2**(7), 460–464 (2006)
29. N. Roch, S. Florens, V. Bouchiat, W. Wernsdorfer, F. Balestro, Quantum phase transition in a single-molecule quantum dot. Nature **453**, 633–637 (2008)
30. A. Kogan, S. Amasha, M.A. Kastner, Photon-Induced Kondo satellites in a Single-Electron transistor. Science **304**(5675), 1293–1295 (2004)
31. C. Buizert, A. Oiwa, K. Shibata, K. Hirakawa, S. Tarucha, Kondo universal scaling for a quantum dot coupled to superconducting leads. Phys. Rev. Lett. **99**(13), 136806 (2007)
32. G.D. Scott, D. Natelson, Kondo resonances in molecular devices. ACS Nano **4**(7), 3560–3579 (2010). doi:10.1021/nn100793s
33. L. Kouwenhoven, L. Glazman, Revival of the Kondo effect. Phys. World **14**(1), 33–38 (2001). arXiv:cond-mat/0104100
34. J. Li, W. Schneider, R. Berndt, B. Delley, Kondo scattering observed at a single magnetic impurity. Phys. Rev. Lett. **80**(13), 2893 (1998). doi:10.1103/PhysRevLett.80.2893
35. V. Madhavan, W. Chen, T. Jamneala, M.F. Crommie, N.S. Wingreen, Tunneling into a single magnetic atom: spectroscopic evidence of the Kondo resonance. Science **280**(5363), 567–569 (1998)
36. N. Knorr, M.A. Schneider, L. Diekhöner, P. Wahl, K. Kern, Kondo effect of single Co adatoms on Cu surfaces. Phys. Rev. Lett. **88**(9), 096804 (2002). doi:0.1103/PhysRevLett.88.096804

37. V. Madhavan, W. Chen, T. Jamneala, M.F. Crommie, N.S. Wingreen, Local spectroscopy of a Kondo impurity: Co on Au(111). Phys. Rev. B **64**(16), 165412 (2001). doi:10.1103/PhysRevB.64.165412
38. A. Zhao, Controlling the Kondo effect of an adsorbed magnetic ion through its chemical bonding. Science **309**(5740), 1542–1544 (2005). doi:10.1126/science.1113449
39. L. Gao, W. Ji, Y.B. Hu, Z.H. Cheng, Z.T. Deng, Q. Liu, N. Jiang, X. Lin, W. Guo, S.X. Du, W.A. Hofer, X.C. Xie, H. Gao, Site-Specific Kondo effect at ambient temperatures in Iron-Based molecules. Phys. Rev. Lett. **99**(10) (2007). doi:10.1103/PhysRevLett.99.106402.
40. P. Wahl, L. Diekhöner, G. Wittich, L. Vitali, M.A. Schneider, K. Kern, Kondo effect of molecular complexes at surfaces: Ligand control of the local spin coupling. Phys. Rev. Lett. **95**(16), 166601 (2005). doi:10.1103/PhysRevLett.95.166601
41. N. Néel, J. Kröger, R. Berndt, T.O. Wehling, A.I. Lichtenstein, M.I. Katsnelson, Controlling the Kondo effect in CoCu$_n$ clusters atom by atom. Phys. Rev. Lett. **101**(26), 266803 (2008). doi:10.1103/PhysRevLett.101.266803
42. Y. Fu, S. Ji, X. Chen, X. Ma, R. Wu, C. Wang, W. Duan, X. Qiu, B. Sun, P. Zhang, J. Jia, Q. Xue, Manipulating the Kondo resonance through quantum size effects. Phys. Rev. Lett. **99**(25), 256601 (2007). doi:10.1103/PhysRevLett.99.256601
43. T. Uchihashi, J. Zhang, J. Kröger, R. Berndt, Quantum modulation of the Kondo resonance of Co adatoms on Cu/Co/Cu(100): low-temperature scanning tunneling spectroscopy study. Phys. Rev. B **78**(3), 033402 (2008). doi:10.1103/PhysRevB.78.033402
44. P. Wahl, P. Simon, L. Diekhöner, V.S. Stepanyuk, P. Bruno, M.A. Schneider, K. Kern, Exchange interaction between single magnetic adatoms. Phys. Rev. Lett. **98**(5), 056601 (2007). doi:10.1103/PhysRevLett.98.056601
45. U. Fano, Effects of configuration interaction on intensities and phase shifts. Phys. Rev. **124**(6), 1866 (1961). doi:10.1103/PhysRev.124.1866
46. O. Újsághy, J. Kroha, L. Szunyogh, A. Zawadowski, Theory of the fano resonance in the STM tunneling density of states due to a single Kondo impurity. Phys. Rev. Lett. **85**(12), 2557 (2000). doi:10.1103/PhysRevLett.85.2557
47. J. Merino, O. Gunnarsson, Simple model for scanning tunneling spectroscopy of noble metal surfaces with adsorbed Kondo impurities. Phys. Rev. B **69**(11), 115404 (2004). doi:10.1103/PhysRevB.69.115404

Chapter 4
Adsorption of Metal Phthalocyanines on Ag(100)

Since its invention STM has been widely used to study molecules on surfaces. It proved an ideal tool for this purpose, giving access to both the real space adsorption geometry and electronic structure of molecular adsorbates. CuPc on polycrystalline Silver were among the first molecules studied as early as 1987 [1–3]. Today many more MePc molecules on different metal surfaces have been investigated (Me = Cu [1, 3–6], Co [6–8], Fe [9–16], Ni [12], Pd [17], Zn [18], Mn [19, 20], Sn [21]). In general, single MePc molecules adsorb flat on the surface, and their four isoindole groups (see Fig. 4.1) are imaged as a four lobed cross shape in STM topography. Depending on the character of the filled d states of the metal ion, the center appears either as a dip or as a protrusion [6, 12].

The interplay between molecule-substrate and molecule–molecule interactions leads to the self assembly of MePc into highly ordered clusters or layers. The substrate symmetry in relation to the four fold symmetry of the molecules can lead to ordered molecular domains [9]. Multilayer films show a growth structure with the molecular plane parallel [22] or tilted with respect to the surface plane [9, 10]. Many growth studies on noble metal (111) surfaces have been undertaken on Ag(111): CuPc monolayers [23], FePc up to multilayer [10], FePc sub-monolayer [24, 25], single SnPc to multilayer [21]; on Au(111): MnPc and FePc sub-monolayer [19], CoPc multilayers [26], SnPc multilayers [27], single FePc to monolayer [16] and on Cu(111) single CoPc [8, 28] and up to monolayer structures [29]. For the (100) surface studies are more scarce, to our best knowledge only CoPc [30] and NiPc/CuPc monolayers [31] on Cu(100) have been investigated.

In this chapter we present a systematic investigation of the adsorption of four different types of MePc (Me = Fe, Co, Ni, Cu) on the Ag(100) surface. The evolution of the growth is studied step by step by STM starting from single molecules, dimers, larger clusters up to a complete monolayer, and finally multilayer structures. We find that the adsorption on Ag(100) induces chirality in the otherwise achiral MePc molecules. Interestingly, this is not due to a geometric distortion, but rather an electronic effect. Our detailed study allows us to follow the evolution and transfer of chirality from the single molecule level up to the structural level, found in clusters and

Fig. 4.1 The Lewis structure of a metal phthalocyanine. The ligand consists of four isoindole groups linked by four aza-bridging Nitrogen atoms. The different types of Nitrogen atoms are labeled N_a and N_p. MePc can accommodate a large variety of metal ions. The two Carbon C_a and C_b atoms in the benzene rings are labeled to measure chiral distortions (see Table 4.1)

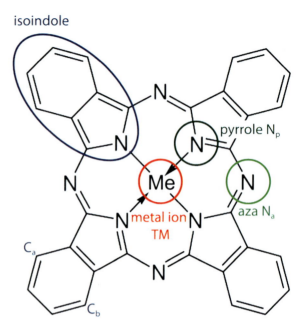

monolayers. The STM results are compared to DFT calculations, which provide further insight into electronic structure of the molecule-substrate complex. The calculations were performed by R. Robles, P. Ordejón, and N. Lorente of the CIN2, Barcelona.

4.1 Chemical Structure of Metal Phthalocyanines

MePc are coordination complexes formed by a central metal ion and an organic ligand. The chemical structure is depicted in Fig. 4.1. The ligand consists of four isoindole groups linked by four aza-bridging Nitrogen atoms. The ligands are arranged around the metal ion in a cross shaped structure. The Me–Pc bond involves charge transfer from the metal to the ligand, leaving the metal in an ionized state, typically $[Me]^{2+}$ for transition metal (TM) ions.

4.2 Adsorption of Single Molecules

4.2.1 Adsorption Configuration

In this first section we will look at the adsorption geometry of single molecules. At first glance the behavior of FePc, CoPc, NiPc and CuPc is very similar, although,

4.2 Adsorption of Single Molecules

as we will see, the contributions to the molecule-surface interaction are different, depending on the 3d level occupation of the central ion.

Figure 4.2 shows a topographic image of low coverage NiPc and CoPc adsorbed on Ag(100). The molecules appear as a four lobed structure, which can be easily identified as the four isoindole groups of the Pc ligand. This indicates adsorption parallel to the surface, as seen in previous investigations [3, 6, 11–13, 28]. Based on their topographic appearance at most bias voltages, the molecules can be further classified in two groups. The TM-center is either a protrusion for FePc and CoPc or a depression for NiPc and CuPc (see Fig. 4.2). This difference is, however, not related to molecular structure, as confirmed by the similar adsorption configuration obtained in DFT calculations (see Table 4.1). Instead, it reflects the high or low degree of coupling between the TM's d electrons near the Fermi level (E_F) and the substrate/tip (see Chap. 5 and [12]).

Two different rotational orientations of the MePc can be identified. By comparing these to atomically resolved images of the Ag substrate, we can quantify the rotation. The ligand axis is rotated by $\pm 30°(\pm 2°)$ with respect to the [011] surface lattice vectors, similar to results found for CuPc on Cu(100) [3]. Moreover, by extrapolating the surface lattice imaged next to a MePc, we find that the metal ion is located over an Ag hollow site (see Fig. 4.2b. The DFT calculations confirm this adsorption site with a 0.14 eV energy gain over the next stable configuration.

Additionally, DFT identifies the driving force for the azimuthal orientation of the molecular axis as the bond between the aza-Nitrogens (N_a) and the underlying Ag atoms. In order to minimize the N_a–Ag distance, the molecule rotates its isoindole axis by 30° with respect to the [011] surface vector. Both theory and experiment find very similar, TM-indepedent, configurations for all MePcs (see Fig. 4.2b, confirming the leading role of N-substrate interactions in determining the molecular orientation.

In the calculations the molecule becomes slightly concave after deposition, meaning that the central ion moves towards the surface and the outer H atoms away from it. In the LDA approximation this amounts to a difference of ~ 0.4 Å. Likewise the presence of the molecule induces a small distortion in the substrate: the Ag atoms beneath the pyrole-Nitrogen (N_p) are shifted ~ 0.1 Å below the average level, while the Ag atoms under the N_a and some of the atoms under the benzene rings lie ~ 0.1 Å above the average. The molecule-substrate distances z, shown in Table 4.1, have been calculated as the difference in the z coordinate between the TM atom and the Ag atom

Table 4.1 Computed height distances z (in Å) between the TM ion and the substrate obtained within different DFT approximations, and height difference between C_A and C_B atoms at the end of the benzene ring (see Fig. 4.2b) obtained with LDA

	LDA	GGA	GGA + vdW	$z(C_A - C_B)$
FePc	2.43	2.76	2.66	0.06
CoPc	2.51	3.08	2.71	0.05
NiPc	2.53	3.59	2.73	0.03
CuPc	2.46	3.47	2.66	0.05

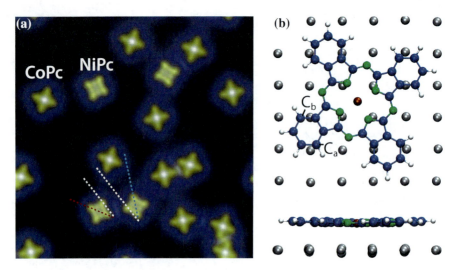

Fig. 4.2 a STM topography of CoPc and NiPc codeposited at room temperature on the Ag(100) surface (0.5 nA, −1.0 V, 13.4 nm × 13.4 nm). The *blue/red lines* indicate the ±30° rotation of the molecular axis with respect to the [011] surface lattice vectors (*white line*) determined from atomically resolved images of the Ag surface. **b** Adsorption configuration of MePc on Ag(100) as obtained from DFT. *Upper panel*: *top view* showing the hollow adsorption site. *Side view*: Some of the Ag atoms are shifted with respect to the surface plane

below the N_p. The distance obtained with LDA is ∼2.5 Å for all the four molecules. This functional is known to over-bind, only indirectly compensating for vdW forces, which are not included in the calculations.

The GGA+vdW functional provides a way to directly consider vdW forces, and by comparing plain GGA and GGA+vdW, we can further characterize the effect of the interaction between the TM ion and the substrate. In plain GGA (see Table 4.1), both FePc and CoPc are significantly closer (∼0.5–0.8 Å) to the surface than NiPc and CuPc. The stronger TM-Ag interaction of FePc/CoPc is attributed to a direct participation of metal d orbitals in the interaction with the substrate [32, 33]. Once the vdW interaction is introduced all distances equalize independently of the TM. The vdW correction to the total energy can be quantified: for NiPc and CuPc it is roughly 1 eV larger than for FePc. The vdW forces thus compensate the lack of direct interaction via the TM ion and explains the TM-independent results of a binding distance of ∼2.7 Å (Table 4.1). Similar values were found for CoPc adsorbed on Cu(111) using LDA and GGA+vdW methods [34]. On less reactive Au(111) surfaces, however, the distances obtained with LDA for different MePcs do not level out and still reflect a TM-dependent behavior [32].

4.2.2 Orbital Specific Electronic Chirality

STM images of NiPc and CuPc present chiral contrast, despite the achiral structure of MePc. The chirality is opposite for the two molecular orientations (+30°, −30°) of NiPc/CuPc (see Fig. 4.3). According to the clockwise and counterclockwise rotation of the ligand axis with respect to the [011] surface vector, these enantionmers will be denoted as r (right) and l (left). The chiral contrast depends on the bias used for imaging. At negative voltages it is clearly visible and progressively disappears for positive values, as shown in Fig. 4.4. Such chiral contrast is only observed for CuPc and NiPc, not for CoPc and FePc. This behavior suggests that chirality in this system is mainly of electronic origin, i.e. not related to the molecule conformation.

Both the molecule and the substrate posses four-fold symmetry. The origin of the chirality lies in the misalignment of surface and molecular symmetry axes (see Fig. 4.2). As has been observed in other systems [35–37], this induces asymmetric molecule metal interactions that can lead to chiral distortions of the molecule. However structural distortions cannot not explain the observed voltage dependency of the chiral contrast.

DFT calculations reveal more about the mechanisms behind this electronic chiral contrast. We quantify the contribution of conformational distortions in the chiral contrast. We use the height difference between two opposite C atoms at one of the benzene ring as a measure of their torsion (see Tables 4.1 and 4.2b. We find very small values for all MePcs, even for CoPc and FePc, which do not show any chirality in the STM images. This confirms that the chirality contrast has an electronic origin rather than a conformational one.

The DFT results show a charge transfer of roughly one electron into the molecule, as well as a strong hybridization between molecular and substrate states. The symmetry of electronic structure will therefore be based on the both substrate and molecule. This structural arrangement of molecule + surface is itself chiral due to the rotation of the molecule (Fig. 4.2b, hence leading to "chirally" perturbed molecular frontier orbitals, without any structural distortion.

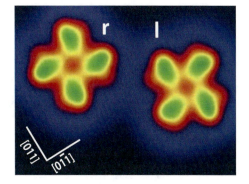

Fig. 4.3 STM topography of CuPc adsorbed on Ag(100). Molecules rotated by ±30° with respect to the [011] crystallographic direction show r and l chirality. (0.1 nA, −0.3 V, 5.8 × 4.4 nm)

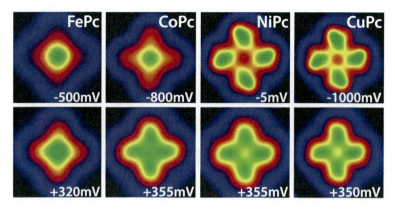

Fig. 4.4 Voltage dependent topographies of four MePcs. For CuPc and NiPc negative voltage images with maximum chiral contrast are displayed. The appearance of CoPc and FePc does not vary in the range of negative voltages investigated, thus representative images are displayed. For positive voltage, all MePcs appear achiral for the studied voltage range <1000 mV (0.1nA, 3 × 3nm)

This electronic chirality is imprinted to each molecular orbital (MO) to a different degree, depending on their spatial distribution and interaction with the substrate. The effect is mainly observed in orbitals exhibiting a nodal plane in the ligand axis (see Fig. 4.5). The a_{1u} orbital (~gas phase HOMO) exhibits the strongest distortion, promoted by its double lobe structure at the benzene ring. The chiral contrast at the $2e_g$ state (~gas phase LUMO), also with nodal planes at the ligand axis, varies with

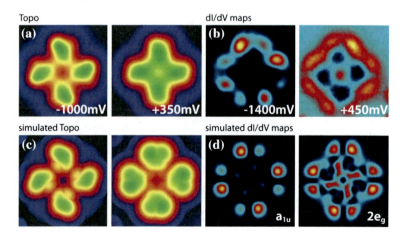

Fig. 4.5 Origin of the chiral contrast: **a** Voltage dependent topographic images of CuPc **b** dI/dV maps taken at the positions of the a_{1u} (~gas phase HOMO) and the $2e_g$ orbitals (~gas phase LUMO) **c** calculated STM images using DFT corresponding to the topographies in (**a**). **d** Theoretical conductance maps at the peak position of the a_{1u} and $2e_g$ orbitals

energy and is strongly reduced at positive bias. TM d states and MO without such nodal planes do not exhibit any significant chirality. This orbital specificity leads to a bias voltage dependent chiral appearance of the MePcs depending only on the particular electronic structure of the molecule in question (see Fig. 4.4). The CuPc and NiPc appear strongly chiral at negative bias, due to the presence of asymmetric orbitals, namely the a_{1u} and the partially occupied $2e_g$ states (see Chap. 5, page 73). In contrast at positive bias the tunneling occurs through the more symmetric $2e_g$ orbital resonance, hence no chirality is observed. TM d orbitals and other interface related resonances with maximum intensity at the ligand hinder the observation of chirality for CoPc and FePc (see Fig. 4.4). Note that the dI/dV maps of the CoPc's a_{1u} orbital do indeed confirm its chirality, as can be seen in the map taken at -1500 mV in Fig. 5.3 on page 79.

4.3 Monolayer Growth

We have seen that single MePc molecules adsorbed on Ag(100) have a chiral electronic structure. A key question is whether the chirality of the individual molecules can be transferred to supramolecular organizational structures. For that to occur each enantiomer must form domains consisting only of one handedness. Chiral molecules have intrinsically some kind of stereospecific interaction (hydrogen bonds [38, 39], or lateral vdW forces), facilitating the formation of supramolecular chiral domains. In the case of achiral molecules the selectivity required can be provided by the interaction with the substrate, which makes it possible to form chiral clusters. This is done by asymmetric molecular interactions due to a mismatch between molecule and substrate symmetry axis, and/or lattice constraints [36, 37].

We studied the evolution of supramolecular structures for the CuPc/Ag(100) system by investigating higher coverages from dimer up to the complete monolayer. These results are contrasted to CoPc layers to determine possible effects of the central metal ion.

4.3.1 Evolution of Chirality in CuPc and CoPc Structures

STM topography images with increasing coverage of CuPc illustrate the evolution of the organizational chirality during the self-assembly process (see Fig. 4.6). The mechanisms behind each step are schematically illustrated in Fig. 4.7.

As a first observation, we notice that CuPc dimers preferentially form with molecules of the same handedness, while mixed configurations with r and l molecules are rare. This is the signature of a chiral recognition process occurring at the single molecule level.

To answer the question whether this chirality is transferred to the supramolecular level, we consider the smallest homochiral structure. The dimer has four possible

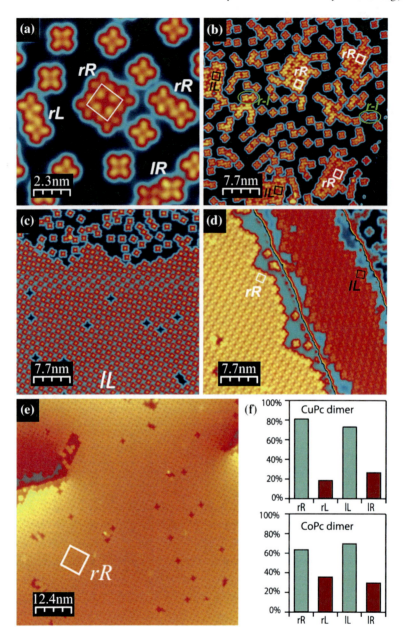

Fig. 4.6 Coverage dependent evolution of organizational chirality: **a** At very low coverages, small clusters nucleate due to attractive intermolecular interactions. Dimers are found in all possible bonding configurations. **b** At intermediate coverage, chiral recognition occurs for larger clusters, leading to a racemic mixture of rR or lL structures. **c** Ostwald ripening favors the growth of one type of domain, leading to spontaneous symmetry breaking. **d** At monolayer completion, chiral purity is achieved on each terrace. **e** A homochiral rR domain extending over a large terrace and crossing a screw dislocation. **f** Probability to find a dimer of each configuration for CuPc (~100 counted dimers) and CoPc (~100 counted dimers). All images are CuPc/Ag(100) except for (**c**) which is CoPc/Ag(100)(0.1nA, −1.0V)

4.3 Monolayer Growth

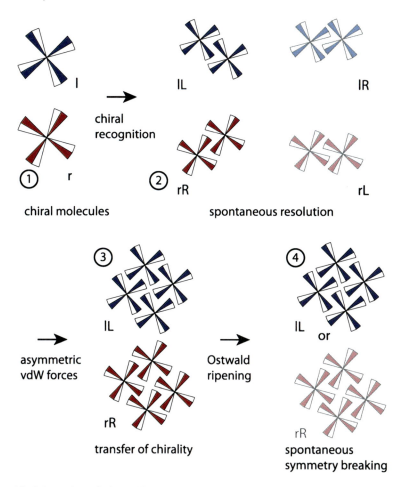

Fig. 4.7 Schematics of the self-assembly mechanisms that induce homochirality on the CuPc/Ag(100) surface. The combination of adsorbate-substrate matching and vdW interactions is strong enough to induce chiral recognition (phase separation of r and l molecules), and transfer the chirality from the single-molecule level to the organizational level (r/l forming only rR/lL islands). Ostwald ripening combined with reversible single-molecule chirality favors the homochiral growth of the largest cluster at the expense of the smaller ones, leading to spontaneous 100% enantiomeric excess on each terrace

bonding configurations, which express its two levels of chirality: the individual determined by the orientation of the molecules (r or l) and the organizational, meaning the position of the molecules relative to each other. The ligand axis of one molecule can either be placed on the right or on the left of the other. These structural orientations will be denoted with capital letters R and L, depending on the rotation of the superlattice with respect to the molecular axis (see Fig. 4.8 on page 60). Consequently, the four structures are rR, rL, lL, lR (see Figs. 4.6a and 4.7). When we quantify the probability to encounter a certain type of dimer, it becomes clear that they are not

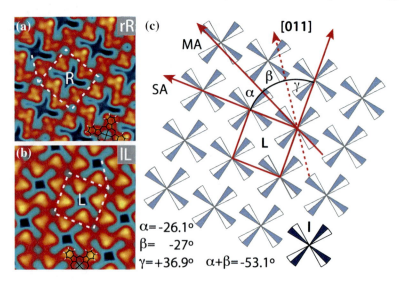

Fig. 4.8 High resolution STM topographies of the chiral clusters **a** rR and **b** lL domains recorded with $V_b = +0.52\,\text{V}$ and $-0.10\,\text{V}$ respectively (0.1 nA, 4.2 × 4.2 nm). **c** The lL superlattice configuration. The molecular axis (MA) and superlattice unit cell directions (SA) and angles with respect to the [011] surface direction are indicated

evenly distributed: type rR and type lL dimers occur far more often than lR or rL (see Fig. 4.6f). Hence a transfer of chirality from the single to supramolecular level is already present for dimers favoring rR and lL arrangements. The energy difference between these configurations, which is due to the interaction with a single neighbor, is not yet large enough to completely suppress the formation of the metastable rL/lR structures. Once the number of nearest neighbors (NN) is larger than one the suppression is complete. For tetramers structures with two NN and larger clusters with four NN, only rR and lL clusters are observed (Fig. 4.6b). These clusters form a racemic mixture, consistent with the fact that these structures are energetically equivalent.

The behavior of CoPc is very similar, despite the non-chiral topography appearance. All clusters with NN > 1 are found in the rR or lL configurations. In dimer structures, however, the general trend of preferring rR and lL, is less pronounced (see Fig. 4.6f). This might either be an empirical problem, as the total number of counted dimers was less than in the CuPc case, or due to the weaker vdW interaction (see Table 4.1) and/or $2e_g$ hybridization (see Sect. 5.2 on page 76) with the substrate in CoPc compared to CuPc. This could allow slight rotation of the former in order to minimize the energy cost of the less favorable configurations (lR or rL). vdW interactions do not only affect the molecule-substrate separation, but also the self assembly, leading to similar organizational structures for the different MePc.

At coverages above approximately 0.5 ML, large islands of either rR or lL type develop on each terrace (Fig. 4.6), growing at the expenses of the smaller ones. Near the completion of the monolayer, each terrace consists of only one enantiopure domain phase (Fig. 4.6d). This Ostwald ripening process occurs due to the continuous exchange between the molecular clusters and a two-dimensional "gas" of adsorbed molecules, similarly to what is known to occur for metal systems [40]. It amplifies small initial fluctuations in the population of rR and lL enantiomers.

For this homochiral growth mechanism it is essential that the molecules can switch chirality during diffusion by rotating over the surface. This is made possible by the achiral nature of the molecules, and their high mobility on surface, when deposited at room temperature. The low energy barrier for a molecule to switch chirality, i.e. rotate, is also reflected in tip induced diffusion experiments, where the molecule repeatedly changes the azimuthal adsorption angle during the lateral diffusion.

Mechanisms that enhance random imbalances in chirality during crystallization are essential to understanding chiral selection processes that take place in nature. In general complete spontaneous symmetry breaking is a rare process for molecules to occur in solution [41, 42]. On surfaces the constrained growth in two dimensions can be used to induce spontaneous chiral growth. However, usually only a racemic mixture of domains is obtained [13, 43, 44]. Enantiopure molecular layers require an external driving force, such as enantiomeric excess [41], a chiral modifier [42], chiral solvent [45], or magnetic field [46]. Here we presented a case where symmetry breaking of chiral molecular layers occurs spontaneously, and is complete at the single terrace level. The extension of the homochiral layers is thus limited only by the morphology of the substrate, not by kinetic or thermodynamic effects.

4.3.2 Supramolecular Structure

We have determined the supramolecular structure of the self assembled CuPc monolayer/cluster by STM. Figure 4.8 show two high resolution STM images of the to observed chiral domains.

Let us consider these structures in more detail for the case of the l enantiomer. The same arguments hold true for the r type. The superlattice structure that we observe for the l molecule, shown in Fig. 4.8c, is lL, which is commensurate with the surface with a 5×5 periodicity. We find that in the superlattices the angle between molecular and surface symmetry axes reduces slightly from $\beta = -30$ to $-27°$ as compared to the single molecules (negative sign corresponds to counterclockwise, i.e. left, rotation). A similar reduction of the azimuthal angle is also observed in the relaxed structures obtained in the calculations.

The lL superlattice consists of a square unit cell, rotated by $\alpha = -26.1°$ with respect to the molecular axis direction, meaning $\alpha + \beta = -27° - 26.1° = \gamma = -53.1°$ rotation with respect to the surface primitive vectors. Due to the 4-fold symmetry of the surface this is equivalent to a $-53.1° + 90° = +36.9°$ rotation, which defines the rotation angle in the reference frame of the surface. In Wood's notation, the lL superstructure would then be written as $5 \times 5R37°$, whereas in matrix notation, using the basis of the surface unit cell, the same structure is expressed as

$$M_{lL} = \begin{pmatrix} 4 & 3 \\ -3 & 4 \end{pmatrix} \quad (4.1)$$

Maintaining the nomenclature for the dimer, we use the letter L to denote its chirality in reference to the counterclockwise (left) rotation of the superlattice unit cell with respect to the molecular axis.

The corresponding mirror symmetric superstructure (lR) would be rotated by $+26.1°$ with respect to the molecular axis, leading to a unit cell closely aligned ($-27° + 26.1° = -0.9°$) with respect to the primitive vectors of the Ag(100) surface. The equivalent superlattice matrix is:

$$M_{lR} = \begin{pmatrix} 5.001 & 0.087 \\ -0.087 & 5.001 \end{pmatrix} \quad (4.2)$$

Evidently, this structure is not commensurate with the surface. Because the molecule-substrate interaction dominates over intermolecular forces, the molecules accommodate to the surface lattice, forming a slightly modified 5×5 superstructure. The closest commensurate superlattice corresponds to a rotation of the unit cell by $+27°$ with respect to the molecular axis, hence a $5 \times 5R0°$ structure or:

$$M_{lR} = \begin{pmatrix} 5 & 0 \\ 0 & 5 \end{pmatrix} \quad (4.3)$$

We have thus found the four possible superlattice structures: for the l enantiomer lL $5 \times 5R37°$ respectively lR $5 \times 5R0°$ and for the r orientation $5 \times 5R37°$ for rR and $5 \times 5R37°$ for rL. Note that the latter structure coincides with lR, however the molecules are a different enantiomer, i.e. rotated in the opposite direction.

In the low temperature STM images only the most stable configurations lL and rR were observed, as the system was slowly cooled. By fast freezing small rL/lR clusters were occasionally created (Data not shown). In low energy electron diffraction (LEED) experiments at room temperature, all of these four domains are found to be coexisting. The broad spot at 0° in Fig. 4.9 indicate the presence of rL/lR clusters

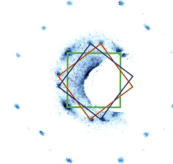

Fig. 4.9 LEED pattern for a submonolayer coverage CuPc/Ag(100) recorded at room temperature showing the coexistence of rR (*orange*), lL (*blue*), and rL/lR domains (*green square*)

4.3 Monolayer Growth

with short range order. The fact that all of them are presences suggest a small energy difference between rL/lR and lL/rR cluster, which confirmed in the next section by DFT.

4.3.3 Origin of the Transfer of Chirality

The energetics and mechanisms involved in the formation of chiral supermolecular clusters can be understood through DFT calculations for the lL and lR structures. Two possible pathways, which have been found to cause chirality in molecular clusters on surfaces were considered: Substrate mediated interactions and direct vdW interaction between molecules.

4.3.3.1 Substrate-Mediated Molecular Interactions

Substrate mediated interactions can lead to chiral recognition due to long-range Coulomb forces combined with adsorption site constraints [47, 48]. Previous theoretical investigations have shown this to occur for molecules with proper gas-phase chiral conformation, i.e. tartaric acid adsorbed on Ni(100) [49] and phenylglycine on Cu(110) [47, 48]. To investigate the role of these forces for MePc on Ag(100), our collaborators P. Ordejón and N. Lorente have modeled the adsorption of single CuPc molecules in a huge 15×15 supercell to avoid possible periodic boundary effects using the SIESTA code. Indeed the electron density of the Ag atoms surrounding the molecule show a chiral perturbation of the metal states resulting in a distortion of the electrostatic potential around the molecule (Fig. 4.10).

The potential is asymmetric up to a distance of about $15\,\text{Å}$ from the Cu ion, which is larger than the distance to the benzene ring of the nearest neighboring molecule ($13\,\text{Å}$), suggesting a possible influence in the chiral recognition process. We integrated the induced electrostatic potential at the position that would be occupied by a second molecule in the lL and lR adsorption sites. The electrostatic interaction energy of one molecule due to the presence of the other one is given by:

$$E_{el} = \int \rho_{tot}(\vec{r}) \cdot V_H(\vec{r}) d\vec{r} \qquad (4.4)$$

where, ρ_{tot} represents the total charge density of the second molecule (sum of electronic and nuclear charges), V_H is the electrostatic potential induced by the first molecule, and $\vec{r}$ the position. Note that, for such large molecules, a fully self-consistent calculation where the two molecules are included in the computation of the potential is at present prohibited by the supercell size limitations. After integration over the region occupied by the second molecule we find a negligible energy difference of $\Delta E = E(lL) - E(lR) = 3.5\,\text{meV}$. Our ΔE is about one order of magnitude smaller than the substrate-induced interaction estimated for opposite phenylglycine

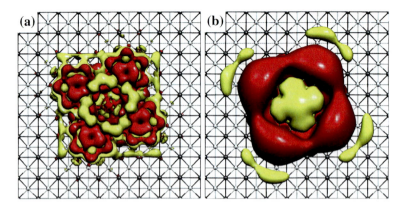

Fig. 4.10 a Extended differential electron density and **b** electrostatic potential of an l oriented CuPc molecule adsorbed on a 15 × 15 supercell (only partially shown). The contours correspond to values of $\pm 7 \times 10^{-4}\,\text{e}/\text{Å}^3$ and $\pm 0.07\,\text{eV}$ for the density and potential, respectively. *Yellow* indicates positive, *red* negative values

enantiomers and adenine on Cu(110) [47, 48]. Moreover these molecules have a much smaller surface footprint compared to MePc, suggesting that the effect of substrate-mediated interactions between enantiomers depends not only on charge transfer but also on the size of the adsorbates. In conclusion due to the small energy difference, we can exclude substrate-mediated molecular interactions as the driving force behind the chiral recognition process.

4.3.3.2 Intermolecular Van-der-Waals Forces

The second pathway we considered were intermolecular van-der-Waals forces. To compute the contributions to the chiral recognition, two different functionals were used, following the same arguments used for the adsorption of single molecules (see page 54). LDA reproduces the TM-substrate interactions quite well, however weak interactions like vdW are only indirectly described. The second functional DSRLL is designed specifically to work with vdW forces and hence should provide a more accurate description. Using both functionals, the relaxed structures and energies for the two configurations (lL and lR) were obtained.

The structure of the monolayer is mainly determined by the interaction between each individual molecule and the Ag(100) surface. The angle between the molecular symmetry axes and the substrate crystallographic directions in the relaxed structures is essentially the same for both structures (see Fig. 4.11), indicating that intermolecular interactions in the monolayer are less important than molecule-surface interactions. This is the case for both of the DFT functionals used.

However, the contact distances between neighboring molecules are different for the lL and lR structures. This is due to the fact that both assemblies have the same angle between molecule and substrate axes and the same 5 × 5 superlattice peri-

4.3 Monolayer Growth

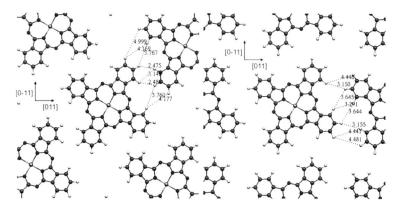

Fig. 4.11 Scheme of the structure of the IL (*left*) and IR (*right*) clusters (Ag atoms are not shown for the sake of clarity), the contact distances between neighboring molecules are indicated. Results obtained with the DSRLL functional

Table 4.2 Energy differences between the IL and the IR structures $\Delta E = E(IL) - E(IR)$ in meV, computed with different DFT functionals

	LDA	DSRLL	DSRLL no vdW
ΔE(ML on Ag100)	−69	−41	3
ΔE(free ML)	−94	−74	16

Negative values indicate a preference for the IL phase

odicity, whereas the unit cell vectors are different. As a consequence, neighboring molecules lie in different directions. The intermolecular contacts are shorter for the monolayer in the IL phase, as shown in Fig. 4.11, because the relative orientation between molecules is such that two phenyl groups are closer in this case.

Finally in Table 4.2 the energy differences between the IL and IR phases are shown. If the vdW contributions are not considered (DSRLL no vdW) both structures are energetically almost equal. This means that (i) the energy of interaction between the molecule and the surface is similar in both structures, due to the similarity in the angles discussed previously; (ii) the molecules are sufficiently far so that the repulsive part of the interactions (which is present in the calculation) is negligible, and (iii) the effect of the interaction between molecules through the electronic states of the metal does not play an important role in the energy difference between the phases.

However, once vdW forces are included, the two chiral structures become separated by 41 meV, which clearly identifies vdW interactions as the driving force for the chiral interaction. These (attractive) vdW forces between neighboring molecules are stronger in the IL phase due to the slightly closer distances between the molecular endgroups.

It is interesting to note that the energy difference between both phases is smaller for the monolayer over the metallic surface than for the free monolayer, by almost a factor of two. This may be due to a screening of the vdW interactions by the metallic surface [50], hereby reducing the effect on the total energies as well as the energy difference. The LDA results show the same tendencies as those obtained by DSRLL, albeit with larger energy differences. This is a known shortcoming; LDA tends to overestimate the binding at short distances in vdW complexes.

4.4 Multilayer Growth

We investigated the growth behavior of CuPc beyond the first monolayer. In order to do this the sample preparation differs slightly. The first monolayer was deposited at room temperature. It acts as a wetting layer and completely covers the surface of the sample in the manner described in Sect. 4.3.1. We then deposited the following layers at 77 K in order to grow stacks of up to 5 molecular layers in a Stranski-Krastanov growth mode, as seen in Fig. 4.12. A similar growth mode has been observed for SnPc on Au(111) [27] and FePc on Ag(111) [10].

The adsorption configuration is different in each layer (see Fig. 4.13). The molecules of the second layer lie flat on the first layer molecules, with the central Cu ions still aligned, and the ligand axis azimuthally rotated by 45°, as has been seen for CoPc multilayers [22]. On the third layer, flat lying molecules coexist with ones slightly tilted out of the surface plane. The tilt can be roughly estimated from the height differences in topographic images $\beta \approx 2°$. Additionally and independently of the tilt, part of the molecules experience a shift of the position of the central TM ion, $\alpha \approx 43° \pm 10°$. On the 4th and 5th layer all molecules are tilted by $\beta \approx 2.5°$ and $\beta \approx 3°$ respectively as well as shifted. These value agrees the observations on CoPc multilayers on Au(111), which found $\beta = 3°$ for the second layer and $\beta = 4°$ for the third layer [26].

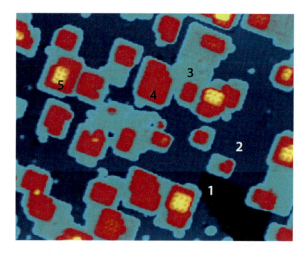

Fig. 4.12 STM topography of CuPc multilayers on Ag(100). The different molecular layers are marked 1−5 (0.01 nA, +1.5 V, 77 × 49 nm)

4.4 Multilayer Growth

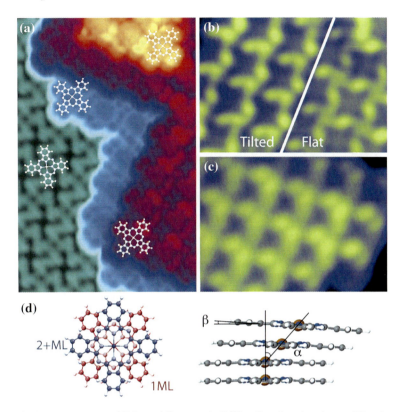

Fig. 4.13 STM topography of CuPc multilayer on Ag(100): **a** Pseudo-colored zoom. The adsorption geometry of molecules within the layers is indicated (0.02nA, +1.5 V, 14 × 9 nm) **b** Topographic zoom of the third layer showing tilted and flat adsorption geometries **c** Topographic zoom of fourth layer, here all molecules are tilted. **d** Schematics of the multilayer adsorption: starting form the second layer the ligand axis is rotated by 45°, in higher layers the molecules are tilted and off-centered

The tilted/shifted adsorption geometry is driven by the $\pi - \pi$ interaction between neighboring molecules, which becomes increasingly stronger due to weaker influence of the substrate-molecule interaction. Ultimately the configuration found in bulk MePc crystals is reached. This behavior has been observed in several other multilayer MePc systems, however depending on the strength of molecule substrate interactions it occurs already in the second layer for FePc/Ag(111) [10], CoPc/Au(111) [26] and on the first layer on NaCl [27]. The α angle lies also roughly within the range of previously observed values of $\alpha \approx 30°$ [22].

4.5 Summary

We studied the adsorption with varying coverage of four MePcs (Me = Fe, Co, Ni, Cu) on Ag(100) using scanning tunneling microscopy and comparing them to ab-initio calculations.

Single molecule adsorption All four single MePc molecules independent of the TM show the same configuration on Ag(100). The π interaction between the macrocycle and the substrate leads to a flat adsorption geometry with the center metal ion situated on a Ag hollow site. The bond optimization between the aza-Nitrogen and the Ag substrate atoms results in two possible orientations of the molecular axis rotated by $\pm 30°$ with respect to the [011] surface vector.

Single molecule chirality The molecular adsorption configuration is chiral, due to the mismatch between molecular and surface symmetry axis. This imprints chirality to the frontier molecular orbitals without perturbing the structural conformation of the molecule. The degree of distortion of the MO depends on their symmetry and spatial distribution. The a_{1u} orbital and the $2e_g$ are the most affected, the former more so. This orbital specific chirality can be disentangled in dI/dV maps. This can be observed as chiral contrast in topography images for NiPc/CuPc at negative bias, because the only contribution to the tunneling current comes from chiral orbitals (a_{1u}). For FePc/CoPc the existence of a filled achiral MO hinders this contrast.

Monolayer growth-transfer of chirality We studied higher coverages of CuPc and CoPc, to investigate the self assembly. The molecules grow in clusters, with a 5×5 periodicity. Further we find that the molecules express chirality on a supramolecular level. At low temperatures we observe two different chiral domains. We find that within MePc clusters asymmetric vdW interactions lead to the transfer of chirality from single molecule level to the organizational. The origin of the asymmetry lies in the mismatch between molecule and substrate symmetry axis in combination with a lattice matching constraint imposed by molecule-substrate interactions. Because these interactions are vdW, the transfer of chirality does not dependent on the TM ion. This mechanism should hold true for all four MePc. The strength of these vdW forces depends on the number of neighboring molecules. In CuPc and CoPc dimer structures the transfer of chirality is not yet perfect, however once a second neighbor is added it becomes univocal.

Homochirality We observe homochiral growth for CuPc monolayers, delimited by surface steps. This spontaneous symmetry breaking is based on the Ostwald ripening of smaller clusters, made possible by the thermally induced molecular diffusions and rotations that switch the chirality of single MePcs during the deposition at room temperature. Since vdW interactions are the driving force, this result should be general for all four MePc.

Multilayer adsorption By first depositing a wetting layer of MePc at RT and subsequent deposition of additional layers at 77 K, we have studied the evolution of self assembly of up to five multilayer of CuPc on Ag(100). The molecules grow in a Stranski-Krastanov growth mode. On the second layer the molecular axis is rotated by $45°$ while maintaining the alignment of the TM ions between the first and second layer. A gradual tilting from the surface plane occurs. On the third layer molecules that are slightly tilted coexist with ones in flat configurations, while on the higher layer all molecules are tilted. The gradual tilting reflects how, in absence of stronger molecule-substrate interaction, the molecules prefer to organize in the closed packed structure observed in bulk films, with vertical stacks formed by $\pi-\pi$ interactions.

References

1. J. Gimzewski, E. Stoll, R. Schlittler, Scanning tunneling microscopy of individual molecules of copper phthalocyanine adsorbed on polycrystalline silver surfaces. Surf. Sci. **181**(1–2), 267–277 (1987). doi:10.1016/0039-6028(87)90167-1
2. H. Ohtani, R.J. Wilson, S. Chiang, C.M. Mate, Scanning tunneling microscopy observations of benzene molecules on the Rh(111) $-$ (3 × 3)($C_6H_6 + 2CO$) surface. Phys. Rev. Lett. **60**(23), 2398–2401 (June 1988). doi:10.1103/PhysRevLett.60.2398
3. P.H. Lippel, R.J. Wilson, M.D. Miller, C. Wöll, S. Chiang, High-resolution imaging of copper-phthalocyanine by scanning-tunneling microscopy. Phys. Rev. Lett. **62**(2), 171–174 (1989). doi:10.1103/PhysRevLett.62.171
4. G.V. Nazin, X.H. Qiu, W. Ho, Visualization and spectroscopy of a metal-molecule-metal bridge. Science **302**(5642), 77–81 (2003). doi:10.1126/science.1088971
5. X.W. Tu, G. Mikaelian, W. Ho, Controlling single-molecule negative differential resistance in a double-barrier tunnel junction. Phys. Rev. Lett. **100**(12), 126807 (2008). doi:10.1103/PhysRevLett.100.126807
6. X. Lu, K.W. Hipps, X.D. Wang, U. Mazur, Scanning tunneling microscopy of metal phthalocyanines: the d7 and d9 cases. J. Am. Chem. Soc. **118**(30), 7197–7202 (1996). doi:10.1021/ja960874e
7. A. Zhao, Controlling the kondo effect of an adsorbed magnetic ion through its chemical bonding. Science **309**(5740), 1542–1544 (2005). doi:10.1126/science.1113449
8. C. Iacovita, M. Rastei, B. Heinrich, T. Brumme, J. Kortus, L. Limot, J. Bucher, Visualizing the spin of individual cobalt-phthalocyanine molecules. Phy. Rev. Lett. **101**(11), 2008. doi:10.1103/PhysRevLett.101.116602
9. A. Scarfato, S. Chang, S. Kuck, J. Brede, G. Hoffmann, R. Wiesendanger, Scanning tunneling microscope study of iron(II) phthalocyanine growth on metals and insulating surfaces. Surf. Sci. **602**(3), 677–683 (2008). doi:16/j.susc.2007.11.011
10. T.G. Gopakumar, T. Brumme, J. Kröger, C. Toher, G. Cuniberti, R. Berndt, Coverage-driven electronic decoupling of Fe-phthalocyanine from a Ag(111) substrate. J. Phys. Chem. C **115**(24), 12173–12179 (2011). doi:10.1021/jp2038619
11. L. Gao, W. Ji, Y.B. Hu, Z.H. Cheng, Z.T. Deng, Q. Liu, N. Jiang, X. Lin, W. Guo, S.X. Du, W.A. Hofer, X.C. Xie, H. Gao, Site-specific kondo effect at ambient temperatures in Iron-based molecules. Phys. Rev. Lett. **99**(10), (2007). doi:10.1103/PhysRevLett.99.106402
12. X. Lu, K.W. Hipps, Scanning tunneling microscopy of metal phthalocyanines: d6 and d8 cases. J. Phys. Chem. B **101**(27), 5391–5396 (1997). doi:10.1021/jp9707448
13. Z.H. Cheng, L. Gao, Z.T. Deng, Q. Liu, N. Jiang, X. Lin, X.B. He, S.X. Du, H. Gao, Epitaxial growth of iron phthalocyanine at the initial stage on Au(111) surface. J. Phys. Chem. C **111**(6), 2656–2660 (2007). doi:10.1021/jp0660738
14. J. Åhlund, J. Schnadt, K. Nilson, E. Göthelid, J. Schiessling, F. Besenbacher, N. Mårtensson, C. Puglia, The adsorption of iron phthalocyanine on graphite: a scanning tunnelling microscopy study. Surf. Sci. **601**(17), 3661–3667 (2007). doi:10.1016/j.susc.2007.06.008
15. N. Tsukahara, Adsorption-induced switching of magnetic anisotropy in a single Iron(II) phthalocyanine molecule on an oxidized Cu(110) surface. Phys. Rev. Lett. **102**(16), 167203 (2009). doi:10.1103/PhysRevLett.102.167203

16. N. Tsukahara, S. Shiraki, S. Itou, N. Ohta, N. Takagi, M. Kawai, Evolution of kondo resonance from a single impurity molecule to the two-dimensional lattice. Phys. Rev. Lett. **106**(18), 187201 (2011). doi:10.1103/PhysRevLett.106.187201
17. T.G. Gopakumar, M. Lackinger, M. Hackert, F. Müller, M. Hietschold, Adsorption of palladium phthalocyanine on graphite: STM and LEED study. J. Phys. Chem. B **108**(23), 7839–7843 (2004). doi:10.1021/jp037751i
18. M. Koudia, M. Abel, C. Maurel, A. Bliek, D. Catalin, M. Mossoyan, J. Mossoyan, L. Porte, Influence of chlorine substitution on the self-assembly of zinc phthalocyanine. J. Phys. Chem. B **110**(20), 10058–10062 (2006). doi:10.1021/jp0571980
19. Y.H. Jiang, W.D. Xiao, L.W. Liu, L.Z. Zhang, J.C. Lian, K. Yang, S.X. Du, H. Gao, Self-assembly of metal phthalocyanines on Pb(111) and Au(111) surfaces at submonolayer coverage. J. Phys. Chem. C **115**(44), 21750–21754 (2011). doi:10.1021/jp203462f
20. K.J. Franke, G. Schulze, J.I. Pascual, Competition of superconducting phenomena and Kondo screening at the nanoscale. Science **332**(6032), 940–944 (2011). doi:10.1126/science.1202204
21. Y. Wang, J. Kröger, R. Berndt, W. Hofer, Structural and electronic properties of ultrathin tin-phthalocyanine films on Ag(111) at the single-molecule level. Angewandte Chem. Int. Edn. **48**(7), 1261–1265 (2009). doi:10.1002/anie.200803305
22. X. Chen, Y. Fu, S. Ji, T. Zhang, P. Cheng, X. Ma, X. Zou, W. Duan, J. Jia, Q. Xue, Probing superexchange interaction in molecular magnets by spin-flip spectroscopy and microscopy. Phys. Rev. Lett. **101**(19), (2008). doi:10.1103/PhysRevLett.101.197208
23. J. Grand, T. Kunstmann, D. Hoffmann, A. Haas, M. Dietsche, J. Seifritz, R. Möller, Epitaxial growth of copper phthalocyanine monolayers on Ag(111). Surf. Sci. **366**(3), 403–414 (1996). doi:10.1016/0039-6028(96)00838-2
24. S.C. Bobaru, E. Salomon, J. Layet, T. Angot, Structural properties of iron phtalocyanines on Ag(111): from the submonolayer to monolayer range. J. Phys. Chem. C **115**(13), 5875–5879 (2011). doi:10.1021/jp111715a
25. T. Takami, C. Carrizales, K. Hipps, Commensurate ordering of iron phthalocyanine on Ag(1 1 1) surface. Surf. Sci. **603**(21), 3201–3204 (2009). doi:10.1016/j.susc.2009.08.029
26. M. Takada, H. Tada, Low temperature scanning tunneling microscopy of phthalocyanine multilayers on Au(1 1 1) surfaces. Chem. Phys. Lett. **392**(1–3), 265–269 (2004). doi:16/j.cplett.2004.04.121
27. Y. Wang, J. Kröger, R. Berndt, H. Tang, Molecular nanocrystals on ultrathin NaCl films on Au(111). J. Am. Chem. Soc. **132**(36), 12546–12547 (2010). doi:10.1021/ja105110d
28. B.W. Heinrich, C. Iacovita, T. Brumme, D. Choi, L. Limot, M.V. Rastei, W.A. Hofer, J. Kortus, J. Bucher, Direct observation of the tunneling channels of a chemisorbed molecule. J. Phys. Chem. Lett. **1**(10), 1517–1523 (2010). doi:10.1021/jz100346a
29. R. Cuadrado, J.I. Cerdá, Y. Wang, G. Xin, R. Berndt, H. Tang, CoPc adsorption on Cu(111): origin of the C4 to C2 symmetry reduction. J. Chem. Phys. **133**(15), 154701-154701-7, 2010. doi:doi:10.1063/1.3502682
30. M. Takada, H. Tada, Direct observation of adsorption-induced electronic states by low-temperature scanning tunneling microscopy. Ultramicroscopy **105**(1–4), 22–25 (2005)
31. G. Dufour, C. Poncey, F. Rochet, H. Roulet, S. Iacobucci, M. Sacchi, F. Yubero, N. Motta, M. Piancastelli, A. Sgarlata, M. De Crescenzi, Metal phthalocyanines (MPc, M=Ni, Cu) on Cu(001) and Si(001) surfaces studied by XPS, XAS and STM. J. Electr. Spectr. Related Phenomena **76**, 219–224 (1995). doi:10.1016/0368-2048(95)02479-4
32. Z. Hu, B. Li, A. Zhao, J. Yang, J.G. Hou, Electronic and magnetic properties of metal phthalocyanines on Au(111) surface: a first-principles study. J. Phys. Chem. C **112**(35), 13650–13655 (2008). doi:10.1021/jp8043048
33. P. Gargiani, M. Angelucci, C. Mariani, M.G. Betti, Metal-phthalocyanine chains on the Au(110) surface: interaction states versus d -metal states occupancy. Phys. Rev. B **81**(8), 2010. doi:10.1103/PhysRevB.81.085412
34. X. Chen, M. Alouani, Effect of metallic surfaces on the electronic structure, magnetism, and transport properties of Co-phthalocyanine molecules. Phys. Rev. B **82**(9), 094443 (2010). doi:10.1103/PhysRevB.82.094443

References

35. M. Parschau, R. Fasel, K. Ernst, O. Gröning, L. Brandenberger, R. Schillinger, T. Greber, A.P. Seitsonen, Y. Wu, J.S. Siegel, Buckybowls on metal surfaces: symmetry mismatch and enantiomorphism of corannulene on Cu(110). Angewandte Chemie International Edition **46**(43), 8258–8261 (2007). doi:10.1002/anie.200700610
36. N.V. Richardson, Adsorption-induced chirality in highly symmetric hydrocarbon molecules: lattice matching to substrates of lower symmetry. New J. Phys. **9**, 395–395 (2007). doi:10.1088/1367-2630/9/10/395
37. M. Schöck, R. Otero, S. Stojkovic, F. Hümmelink, A. Gourdon, I. Stensgaard, C. Joachim, F. Besenbacher, Chiral close-packing of achiral star-shaped molecules on solid surfaces. J. Phys. Chem. B **110**(26), 12835–12838 (2006). doi:10.1021/jp0619437
38. M. Böhringer, K. Morgenstern, W. Schneider, R. Berndt, F. Mauri, A. De Vita, R. Car, Two-dimensional self-assembly of supramolecular clusters and chains. Phys.l Rev. Lett. **83**(2), 324–327 (1999). doi:10.1103/PhysRevLett.83.324
39. J. Weckesser, A. De Vita, J.V. Barth, C. Cai, K. Kern, Mesoscopic correlation of supramolecular chirality in One-dimensional hydrogen-bonded assemblies. Phys. Rev. Lett. **87**(9), 096101 (2001). doi:10.1103/PhysRevLett.87.096101
40. H. Röder, E. Hahn, H. Brune, J. Bucher, K. Kern, Building one- and two-dimensional nanostructures by diffusion-controlled aggregation at surfaces. Nature **366**, 141–143 (1993). doi:10.1038/366141a0
41. C. Viedma, Chiral symmetry breaking during crystallization: complete chiral purity induced by nonlinear autocatalysis and recycling. Phys. Rev. Lett. **94**(6), 065504 (2005). doi:10.1103/PhysRevLett.94.065504
42. N. Petit-Garrido, J. Ignés-Mullol, J. Claret, F. Sagués, Chiral selection by interfacial shearing of self-assembled achiral molecules. Phys. Rev. Lett. **103**(23), 237802 (2009). doi:10.1103/PhysRevLett.103.237802
43. W. Auwärter, A. Weber-Bargioni, A. Riemann, A. Schiffrin, O. Gröning, R. Fasel, J.V. Barth, Self-assembly and conformation of tetrapyridyl-porphyrin molecules on Ag(111). J. Chem. Phys. **124**, 194708 (2006). doi:10.1063/1.2194541
44. D. Écija, M. Trelka, C. Urban, Molecular conformation, organizational chirality, and iron metalation of meso-tetramesitylporphyrins on copper(100). J. Phys. Chem. C **112**(24), 8988–8994 (2008). doi:10.1021/jp801311x
45. R. Fasel, M. Parschau, K. Ernst, Amplification of chirality in two-dimensional enantiomorphous lattices. Nature **439**(7075), 449–452 (2006). doi:10.1038/nature04419
46. M. Parschau, S. Romer, K. Ernst, Induction of homochirality in achiral enantiomorphous monolayers. J. Am. Chem. Soc. **126**(47), 15398–15399 (2004). doi:10.1021/ja044136z
47. S. Blankenburg, W.G. Schmidt, Long-range chiral recognition due to substrate locking and substrate-adsorbate charge transfer. Phys. Rev. Lett. **99**(19), 196107 (2007). doi:10.1103/PhysRevLett.99.196107
48. S. Blankenburg, W.G. Schmidt, Spatial modulation of molecular adsorption energies due to indirect interaction. Phys. Rev. B **78**(23), 233411 (2008). doi:10.1103/PhysRevB.78.233411
49. W. Hofer, V. Humblot, R. Raval, Conveying chirality onto the electronic structure of achiral metals: (R, R)-tartaric acid on nickel. Surf. Sci. **554**(2–3), 141–149 (2004). doi:10.1016/j.susc.2003.12.060
50. J. Mahanty, Screening of the intermolecular van der waals interaction at a metal surface. Phys. Rev. B **35**(8), 4113–4115 (1987). doi:10.1103/PhysRevB.35.4113

Chapter 5
Electronic and Magnetic Properties of MePc on Ag(100)

For numerous technological applications it is necessary to place molecules on a supporting surface or an electrode. However many interesting physical properties of molecules are perturbed when adsorbed on a surface, complicating the reproducible design of hybrid molecular electronic devices. On a metal surface, the electronic interactions distort the ligand field [1], induce charge transfer processes [2–4], and reduce the d-d electron correlation at metal ions by screening and hybridization [3, 4]. The complex interplay between all these processes is far from being understood, as reflected by the intense research carried out on this topic during the last years.

Most studies of the magnetic properties of MePc have focused on MnPc [2, 5], FePc [2, 4, 6, 7], and CoPc [2–4, 8–13]. A general tendency seems to be that the adsorption on a metallic surface reduces or quenches the magnetic moment of the TM ion. On the other hand, systematic studies of different MePcs carried out by photoemission [14, 15] and DFT [2] show a strong influence of the spatial extension of TM d states on the modification of electronic and magnetic properties at the interface. The perturbation is weakest when the d states near the Fermi level are confined in the molecular plane.

Other factors influence the molecule substrate interaction, and with it the molecular magnetism. For instance, two different adsorption sites for FePc on Au(111) have been observed to lead to two different Kondo interactions [6]. The molecular coverage also plays a role. For coverages close to a monolayer, lateral molecule–molecule interactions increase, weakening the coupling to the substrate [16, 17].

Common to these investigations is the assumption that the magnetic properties of MePc at surfaces depends almost exclusively on the ground state of the metal ion, while the ligand part is regarded mainly as a mediator controlling the environment of the magnetic atom. The case of CoPc on Au(111) is an example, where the appearance and disappearance of a Kondo resonance, i.e., the interaction of the magnetic ion with the substrate is controlled by chemically changing the organic ligand [18]. Later on in this chapter we will show that the ligand cannot be regarded only in such a manner, and that it can actively participate in the magnetism of the entire complex. Indeed, a growing number of experiments indicate that charge transfer to/from delocalized π orbitals can infer spin to purely organic complexes [19–24] as well as play a fundamental role in the magnetism and transport properties of metal–organic [13, 25, 26].

In this chapter we will look at the electronic and magnetic structure of four different MePc (Me = Fe, Co, Ni, Cu) deposited on Ag(100), investigating the mechanisms that lead to changes in both the ion and ligand magnetic moments. Scanning tunneling spectroscopy (STS) combined with DFT calculations, performed by R. Robles, R. Korytár and N. Lorente, allows us to explain the role of charge transfer, hybridization, and correlation on the electronic ground state and the magnetic properties.

We will start by reviewing the gas phase electronic structure to better contextualize the effects of the surface. The electronic structure of adsorbed MePc is discussed by comparing STS data with molecular projected density of states (PDOS) obtained from DFT. Afterwards the magnetic properties of MePc are addressed, using STS to measure the intensity and spatial extension of Kondo resonances in the differential conductance, including the coupling to vibrational and magnetic degrees of freedom. On the basis of these measurements combined with DFT results, we are able to propose a magnetic structure.

The end of the chapter will treat the effect of higher coverages on the electronic and magnetic properties, considering small clusters up multilayer structures of CuPc and CoPc. The relative strength of the different hybridization channels plays a critical role in determining the degree of decoupling through these interactions. For CuPc clusters in the submonolayer regime we find a gradual change in the interaction with the substrate based on intermolecular interactions. Finally the multilayer system shows a complete decoupling from the substrate exhibiting typical effects for molecules in a double barrier junction.

5.1 Pristine MePc: Gas Phase Electronic Structure

It is useful to review the gas phase electronic structure of MePc, in order to better understand the effect of the substrate (see Fig. 5.1).

MePcs are metal organic complexes, consisting of an organic ligand (Pc) and a metal atom in the center of 4 pyrrole N (see Fig. 4.1 on p. 52). For transition metal (TM) centers the bonding to the ligand has an ionic character and the ion is in a $[TM]^{2+}$ state. The molecule has a square planar D_{4h} symmetry. Under this ligand field symmetry, the TM's d states transform as b_{2g} (d_{xy}), b_{1g} ($d_{x^2-y^2}$), a_{1g} (d_{z^2}), and e_g ($d_\pi := d_{xz}, d_{yz}$). Depending on the overlap and energy position, they mix in different degree with the $2p$ states of neighboring C and N atoms. The TM-related d states will be classified in two groups according to their orientation with respect to the molecular/substrate plane: the parallel $d_\parallel := b_{2g}$ (d_{xy}) + b_{1g} ($d_{x^2-y^2}$) or perpendicular $d_\perp := a_{1g}$ (d_{z^2}) + e_g (d_π) states.

5.1 Pristine MePc: Gas Phase Electronic Structure

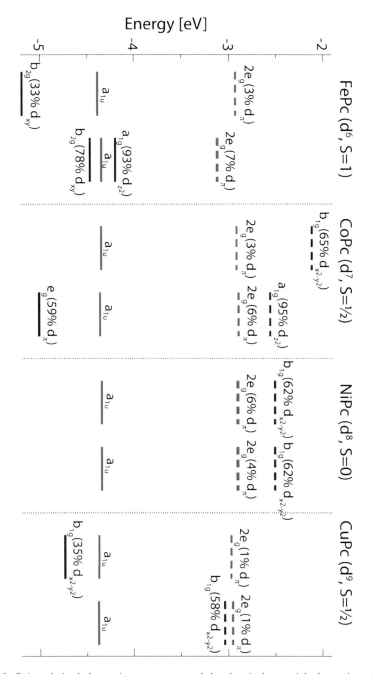

Fig. 5.1 Spin-polarized electronic structure around the chemical potential of gas-phase MePcs: The contribution of different d states is indicated in percentage. *Solid lines* are filled, *dashed lines* empty states. *Blue* indicates states that are largely localized on the Pc ligand part, while *red* represent states that originate mainly from the TM's d states. $a_{1u}/2e_g$ refer to the gas phase HOMO/LUMO states of the Pc ring

The gas phase HOMO/LUMO orbitals of the Pc ring are delocalized π orbitals with marginal TM-d contribution, represented respectively by $a_{1u}/2e_g$. The $2e_g$ (LUMO) is two fold degenerate, without considering the spin degree of freedom.

The TM centers used in this study are Fe, Co, Ni, Cu. All of them are 3d TM with increasing occupation of the d-level from Fe (d^6) to Cu (d^9). Figure 5.1 shows the distribution of the energy levels near the Fermi level obtained with GGA + U, with U−J = 3 eV. The TM-d contribution in each MO is indicated. The resulting magnetic moments are also shown. The electronic structure of the Pc ring is barely affected by the type of TM ion. This is clearly seen for the a_{1u} and $2e_g$ ligand orbitals: The a_{1u} has a negligible d contribution and appears at the same energy for all molecules. The $2e_g$ energy varies slightly with the d configuration of the TM ion, and exhibits some intermixing with the e_g (d_π) state, ranging from 7 % in FePc, to 1 % in CuPc. The different degree of hybridization depends on the proximity of the two levels with e_g symmetry.

The evolution of the electronic configuration of the [TM]$^{2+}$ ions cannot be followed by a simple rigid filling of ligand field split levels, due to the strong d-d correlation and the energy dependent hybridization with ligand states. This is especially true for the complex quasi-degenerate ground state of FePc. The different results obtained within each approximation reflect the discrepancies found in the literature [4, 27–29]. The discussion of these differences is beyond the scope of this work. However, independent of the method used, the total magnetic moment obtained for FePc is always S = 1, which is in good agreement with previous studies [4, 27, 29, 30]. CoPc, with one electron more, presents a S = 1/2($^2A_{1g}$) ground state with a single a_{1g} (d_{z^2}) hole. NiPc, with d^8, is in a closed-shell S = 0 ($^1A_{1g}$) configuration. And lastly the CuPc molecule, with the d^9 Cu ion, has a single hole in the b_{1g} ($d_{x^2-y^2}$) state, resulting in a doublet ground state S = 1/2($^2B_{1g}$).

It is important to point out that in the two S = 1/2 systems, namely CoPc and CuPc, the spin resides in orbitals with very different spatial distribution, i.e., $d_\perp$ and $d_\parallel$ respectively. This will lead to a very different behavior in the interaction with the substrate, as shown in the following sections.

5.2 Single Molecules: Electronic Structure

In this section we present the spectroscopic data of the molecules adsorbed on the Ag(100) surface. The energy of the MOs and other interface-related features is obtained from averaged dI/dV spectra, whereas spatial constant current maps of the dI/dV intensity at specific energies reveals their symmetry and distribution within the molecule. Experimental results are then compared to the density of states projected onto different MOs, which allows the identification of the orbital belonging to the observed peaks, and gives insight into the different processes occurring at the metallic interface.

5.2.1 Spectroscopy of Molecular Orbitals

A series of dI/dV spectra acquired in the energy range of the frontier MOs is displayed in Fig. 5.2a. The a_{1u} and $2e_g$ orbitals of the Pc ring (the gas phase HOMO/LUMO) can be easily identified by studying the spatial distribution of the peaks of the spectra acquired on the benzene (blue), presented as dI/dV maps in Fig. 5.3. The appearance is similar to that found in other substrates [10], except for the chiral imprint discussed in the previous chapter. The energy position of the a_{1u} orbital is similar in all MePcs, continuously varying from -1.14 V in CoPc to -1.40 V in CuPc (unfortunately FePc was not explored in this energy range).

On the other hand, the spectral distribution of the $2e_g$ critically depends on the TM ion: In FePc and CoPc, a single unoccupied peak is observed at $+0.47$ V and $+0.39$ V respectively, whereas CuPc and NiPc exhibit two peaks around the Fermi level, at $-0.29/+0.35$ V and $-0.35/+0.35$ V respectively. Both peaks, however, can be associated with the $2e_g$ due to their similar spatial distribution (see Fig. 5.3a). The localization of the lower energy peak below E_F indicates that the orbital is partially occupied, suggesting a charge transfer from the substrate to the ligand $2e_g$ orbital. The energy splitting of ∼0.65–0.70 eV is within the range of the Coulomb repulsion energies obtained for π-orbitals in aromatic complexes of similar size [19, 31, 32]. We therefore assign the peaks to the single and double occupation of the $2e_g$ state. The sharp peak appearing between the two $2e_g$ peaks is the Kondo resonance (see Chap. 3), which will be discussed in Sect. 5.3.

The MOs with d character can be tracked in the spectra taken at the TM ion (see Fig. 5.2a). The broad peak shifting downwards with increasing d state occupation from FePc (-0.40 V) to NiPc (-1.50 V) is due to the $d_\perp$ states, which are easy to observe in STS because of their strong coupling to the tip. In CuPc, these states lie below the probed energy range. The localization of this peak on the TM ion is confirmed by the dI/dV map of CoPc at -0.51 V shown in Fig. 5.3b.

In CoPc, the $d_\perp$ resonance appears well below E_F, suggesting the complete filling of the a_{1g} (d_{z^2}) orbital, and hence a switch in the charge transfer channel from the ligand $2e_g$ to TM-d states. In FePc, the tail of the $d_\perp$ peak crosses E_F, suggesting that these states are not fully occupied, similarly to FePc on Au(111) [4]. As we will see in the next section this interaction will induce a complex reorganization of charge that involves mixing of $d_\perp$ and ligand orbitals.

The $d_\parallel$ states seem to remain unperturbed after adsorption. Although the orientation parallel to the molecular plane makes them difficult to explore by STM, some features in the spectra of NiPc and CuPc can be tentatively assigned to b_{1g} ($d_{x^2-y^2}$). In the pristine molecules, this orbital is singly occupied in CuPc and unoccupied in NiPc (see Fig. 5.1). The unoccupied contribution is overlapped by a step-like feature that appears around $+0.20$ V in all molecules. However, the higher intensity in this energy range compared to CoPc and FePc hints at the presence of the state in both CuPc and NiPc. The singly occupied b_{1g} orbital in CuPc can be assigned to the small hump observed around -0.40 V, in the same range where the $2e_g$ resonance at the benzene spectra presents a pronounced valley. This valley, absent in NiPc at this energy, may be related to a Fano lineshape originated from a tunneling interference

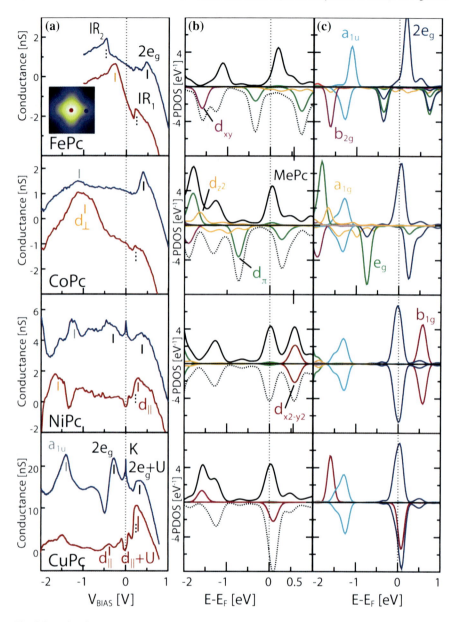

Fig. 5.2 a dI/dV spectra, acquired on the metal ion (*red*) and the benzene ring (*blue*). The latter is shifted for clarity. The labeled features correspond to $2e_g$ (*black*), a_{1u} (*grey*), $d_\perp$ (*orange*), $d_\parallel$ (*green*), interface states (*dotted*). The *inset* shows an STM topography of FePc, where the positions of dI/dV spectra are indicated. Tunneling conditions: 1.0 nA, -1.0 V (FePc, CoPc), 3.0 nA, -2.2 V (NiPc), 3.0 nA, -2.0 V (CuPc). The spectra have been background subtracted. **b** Computed spin-polarized DOS of the MePc (*black*), and its projection onto the TM d states. **c** Computed spin-polarized PDOS projected onto MOs. The PDOS was obtained by GGA + vdW

5.2 Single Molecules: Electronic Structure

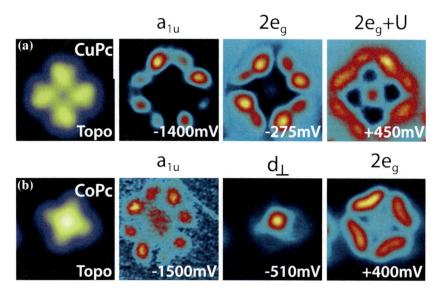

Fig. 5.3 dI/dV maps of **a** CuPc and **b** CoPc, where the $d_\perp$ and ligand a_{1u} and $2e_g$ states can be identified. A topography obtained during the acquisition of maps is displayed on the *left side*. NiPc and FePc present MO distributions similar to CuPc and CoPc respectively

between the occupied b_{1g} and the $2e_g$, analogous to that occurring between resonant and direct channels [33, 34]. Indeed, the dip observed in NiPc at the lower energy side, where $d_\perp$ states overlap with a_{1u}, supports the presence of Fano peaks in these molecules.

Apart from the resonances that maintain the character of pristine MOs, new interface resonances originate from the strong hybridization with Ag electrons, as shown in the dI/dV maps of Fig. 5.4. The step-like IR$_1$ feature at $+0.20$ V, common to all MePcs and only observed on Ag(100), could originate from hybridization with a surface state that is close in energy [35]. In FePc, we observe an additional feature (IR$_2$) below the $d_\perp$ state, which we identify with an interface state with intensity at

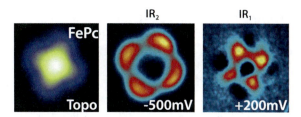

Fig. 5.4 dI/dV maps of interfacial resonances in FePc arising from the hybridization between molecular and substrate electrons (marked as IR$_x$ in Fig. 5.2b). A topography obtained during the acquisition of maps is displayed on the *left side*

the benzene rings, similar to that reported for CoPc on Au(111) [10]. On Ag(100) we also observe a weak resonance, which could be identified with IR_2 at energies between the $d_\perp$ and a_{1u} peaks for CoPc.

We can thus conclude that the charge transfer channel changes depending on the TM. Molecules with frontier $d_\perp$ orbitals (FePc, CoPc) interact through them, although the charge is redistributed on both metal and ligand orbitals. On the other hand when the $d_\perp$ states lie far from the Fermi level (NiPc, CuPc) the interaction occurs through the ligand $2e_g$ MO, which accepts an electron from the substrate.

5.2.2 DFT: Electronic Structure

To correlate the peaks observed in the dI/dV spectra with the molecular electronic structure, we rely on the calculated DOS projected onto the TM-d states and the MOs of the MePc (see Fig. 5.2b, c). In general we find fair agreement between the experimental data and the calculated PDOS, which allows us to assign many of the spectral feature to specific MO, based on the symmetry and spatial distribution of the dI/dV peaks.

The most hybridized states are the a_{1g} orbitals of FePc and CoPc, as expected due to their dominant d_{z^2} character. The TM d_{z^2}-electrons hybridize with the Ag-sp$_z$ states, confirming the strong, direct substrate-TM interaction discussed for the adsorption configuration in Sect. 4.2.1 (p. 52). Additionally, in the case of FePc and CoPc, the minority $2e_g$ and e_g states are mixed together, projecting onto one occupied and one empty spin-down resonance (see Fig. 5.2c). The mixing is induced by the hybridization with the substrate, showing that the picture of slightly distorted MO is no longer valid in this case. This is in line with the appearance of an interface resonance as observed in the STS maps (IR_2 in Fig. 5.4).

For NiPc and CuPc, the hybridization with the substrate turns out to be much smaller, and the PDOS resembles that of the gas-phase molecules of Fig. 5.1. In these two cases the confinement of the b_{1g} orbital in the molecular plane and its σ character reduce the intermixing with the Ag electrons, strongly decreasing the charge transfer to this orbital. Hence, it appears unoccupied in NiPc and singly occupied in CuPc, as in the gas-phase.

The a_{1u} orbital, fixed at about -1.30 eV, is the only MO that is not affected by the d-occupation of the TM ion. In the case of FePc, the spin up and spin down a_{1u} states are exchange-split due to the single-configurational nature of DFT. This broken symmetry reflects the strong spin-polarization of the whole molecule. The position of the $2e_g$ MO, on the other hand, depends on the type of TM ion and reveals a transition in the interaction with the substrate. As in the dI/dV spectra, the PDOS shows that this state is unoccupied in FePc, partially occupied in CoPc, and acquires a charge of approximately one electron in NiPc and CuPc. The splitting between occupied and unoccupied $2e_g$ states is not fully reproduced in the calculations due to the well-known electronic gap problem in DFT, which is especially critical for delocalized π-orbitals [36].

5.3 Single Molecules: Magnetic Structure

In the following we will focus on the analysis of the magnetic structure and its relation to the modified electronic properties studied in the previous section. Experimentally we use the Kondo interaction as an indirect way to probe the molecular spin with a non spin-polarized technique such as STS. For some molecules we observe a complex Kondo behavior, which can be explained through the use of DFT calculations, where the full magnetic structure of the molecules including the ligand and the ion is considered.

5.3.1 Kondo Interaction

The Kondo interaction was introduced in Chap. 3: it is the coupling of a localized impurity spin to a bath of underlying conduction electrons. Its main characteristic is that at temperatures below the characteristic Kondo energy $k_B T_K$, the localized spin is antiferromagnetically screened by the electrons in the Fermi sea, forming a many-body singlet ground state. The many-body state appears as a sharp resonance at E_F, and is detectable by STM. Hence by taking dI/dV spectra around E_F on the molecules, we study their magnetic interaction with the substrate. In the most simple cases the molecular magnetic moment can be extracted from this data.

The dI/dV spectra obtained for the four cases are displayed in Fig. 5.5. FePc and CoPc present featureless spectra, which indicate that the Kondo interaction between these molecules and the Ag(100) surface is either absent or too weak to be observed at 5 K. Relatively flat Co spectra were reported also for CoPc adsorbed on Au(111), for which it was concluded that the filling of the a_{1g} state leads to a complete quenching of the molecular magnetic moment [9]. This interpretation has been recently questioned based on a mixed-valence model of X-ray absorption spectra [4]. In both cases, the quenching of the CoPc magnetic moment appears to be a robust result. On the other hand, FePc does present a Kondo resonance on Au(111), although different interaction strengths have been reported [6, 17]. We attribute the absence of Kondo peaks on Ag(100) to a stronger interaction with the substrate, as reflected by the substantial hybridization of the $d_\perp$ states, which may lead to a mixed-valence configuration of FePc, as discussed later.

In contrast with FePc/CoPc the dI/dV spectra for NiPc and CuPc show many characteristic signatures of low-energy excitations. The most prominent peak, located at zero bias, is the elastic zero bias Kondo resonance K_e (see Sect. 3.4.1 on p. 38). The temperature dependence of its width and intensity are expected to follow expressions derived from Fermi liquid theory [37] and numerical renormalization group (NRG) [38]. We therefore recorded spectra at the benzene at different temperatures. In Fig. 5.6 we display the results obtained for CuPc, confirming the Kondo behavior. Fitting the values of the width (Γ_K) with Eq. 3.18 yields a Kondo temperature of $T_K = 27 \pm 2$ K. The logarithmic behavior of the intensity (G_K) makes it more sensitive to the scattering of the data, but we can see that expressions for both S = 1/2

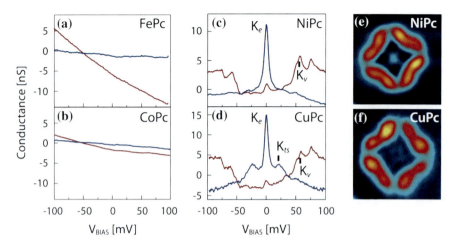

Fig. 5.5 a–d dI/dV spectra acquired around E_F on the metal ion (*red*) and the benzene ring (*blue*). The zero-bias (elastic), vibrational, and triplet–singlet Kondo resonances observed in CuPc and NiPc are labelled as K_e, K_v and K_{ts} respectively. Tunneling conditions: 1.0 nA, −100 mV (FePc, CoPc), 1.1 nA, −100 mV (NiPc), 2.0 nA, −100 mV (CuPc). The spectra have been background subtracted. **e** and **f** d^2I/dV^2 map acquired at −3 and −5 mV respectively, showing the intensity distribution of the elastic Kondo resonance (K_e) of CuPc and NiPc and its resemblance with the dI/dV map of the $2e_g$ orbital. See Chap. 2, p. 27 for a description of the d^2I/dV^2 method

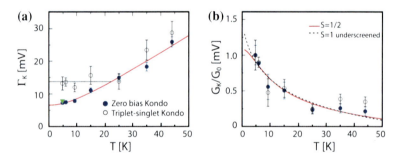

Fig. 5.6 Temperature dependence of the width Γ_K (**a**) and the intensity G_K (**b**) of the elastic Kondo resonance of CuPc/Ag(100), normalized to the intensity at 5 K (G_0). A Kondo temperature of $T_K = 27 \pm 2$ K is obtained by fitting Γ_K with Eq. 3.18 (*solid line*). The (*green*) *triangle* in **a** corresponds to the data obtained for NiPc at 5 K. The *solid* (*dashed*) *line* in **b** is a fit of G_K/G_0 for the S = 1/2 (underscreened S = 1) cases of Eq. 3.19

and the underscreened S = 1 fit reasonably well the data by using $T_K = 27$ K. Unfortunately the two cases diverge only at temperatures that are not experimentally accessible, hence none of the two cases can be excluded based on this data.

Despite of the diamagnetic character of the Ni ion, we find a similar peak at E_F for NiPc. This establishes that the origin of the molecular spin inducing the Kondo resonance is not based on the TM. A Kondo temperature of $T_K = 29$ K can be estimated for this molecule from the width obtained at 5 K (green triangle in

Fig. 5.6a). In addition, the intensity of the resonance in both CuPc/NiPc is not centered at the TM ion, but rather follows the spatial distribution of the $2e_g$ and $2e_g + U$ states, as revealed by the similarity of their dI/dV maps (Figs. 5.3 and 5.5e, f). These two observations make it clear that the interacting spin is indeed localized at the ligand, a result in perfect agreement with our STS observation of a single electron charge transfer to the $2e_g$. Furthermore the experimental values we obtain for the energy of the singly occupied level ($\epsilon = -0.3\,\text{eV}$), its width ($\Gamma \sim 0.30\,\text{eV}$), and the Coulomb repulsion potential ($U = 0.65\,\text{eV}$) confirm that we are indeed in the Kondo regime $|\epsilon|, |\epsilon + U| \geq \Gamma$ (see Chap. 3).

The direct observation of the "Coulomb blockade" peaks and the Kondo resonance allows, unlike previous studies of molecular adsorbates, for a univocal identification of the MO associated to the unpaired spin. As the spin participating to the many-body Kondo state originates from a single orbital, we expect a well-defined value of T_K over the whole molecule, which is confirmed by our observations. This is in contrast with the variations of T_K found for Co porphyrins adsorbed on Cu(111) that suggest a spatially-dependent, multiple orbital origin of the Kondo interaction, possibly due to strong intermixing of TM and ligand orbitals near E_F [26].

5.3.1.1 Coupling of the Kondo Resonance to Vibrational and Spin Excitations

Apart from the Kondo resonance peak, the spectra for NiPc and CuPc show multiple inelastic features. It is well known that inelastic excitations induce step-like increases in the differential conductance spectra symmetrically distributed in energy [39, 40], very much like the ones observed in the TM spectra of Fig. 5.7. In Kondo systems, the coupling of the Kondo state to such excitations leads to additional peaks (K_v and K_{ts}) that appear locked at the same energies [41]. The origin of such side peaks in small aromatic molecules is restrained to vibrational or magnetic excitations, as was discussed in Sect. 3.4.3 on p. 44. The cotunneling mechanisms behind each process are presented in Fig. 5.8.

We can assign the inelastic features found at the TM centers to vibrational excitations by comparing their energies with Raman vibrational modes of the gas-phase CuPc [42], as illustrated in Fig. 5.7a. Raman active modes are the ones with highest intensity in the d^2I/dV^2 spectra due to their symmetrical character with respect to the σ_v planes of the molecule. According to symmetry selection rules, such modes are the only ones that present maximum intensity at the mirror planes of the molecule when the electronic orbitals involved in the inelastic process are symmetrical [43]. This is the case for the b_{1g} and $2e_g$ states. Further we find that all observed modes involve strong TM-N and hence b_{1g}, which is required in order to have a finite inelastic vibrational tunneling to this orbital [43].

Note that the vibrational spectra of NiPc and CuPc are nearly identical, as shown by the twin d^2I/dV^2 spectra displayed in the bottom graph. The step-like and peak-like contributions to the inelastic conductance can be obtained by fitting the dI/dV spectra with step and Lorentzian functions, as shown in Fig. 5.7. From the fit of

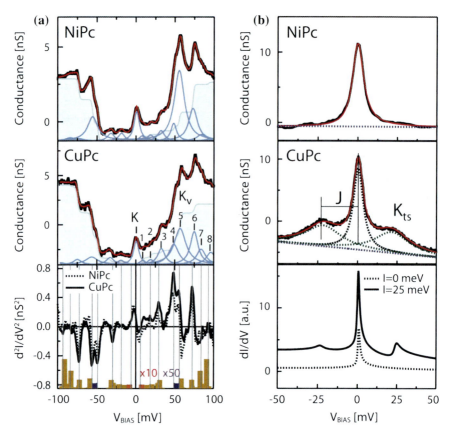

Fig. 5.7 Fit of the background-subtracted dI/dV spectra of NiPc and CuPc measured at the TM ion (**a**) and the benzene ring (**b**), using Fano and Lorentzian functions for the zero and finite bias peaks respectively, and step functions for the inelastic conductance steps observed at the TM ion. The *bottom graph* in **a** is the second derivative (d^2I/dV^2) measured at the TM ion. The energies obtained from the fit are indicated by *dashed lines* in the d^2I/dV^2 spectra. The *bars* correspond to the intensity of calculated Raman modes of gas-phase CuPc [42]. The *bottom graph* in **b** shows the theoretical PDOS obtained for the $2e_g$ close to E_F with a ab initio model, for $I = 25$ meV (*solid*) and 0 meV (*dashed*), see Ref. [44] for more details

the spectra we observe that both NiPc and CuPc present more intense conductance steps at negative bias and more intense Kondo peaks at positive bias. These inverted correlation between vibrational and Kondo features has also been observed for TCNQ molecules adsorbed on Au(111), and has been attributed to the competition between the purely inelastic channel and the one including the Kondo effect [19].

At the benzene ring, in contrast to the spectra at the TM ion, inelastic features appear only for CuPc (shown in Fig. 5.7b). Since both molecules share the same vibrational spectrum the divergence must come from the magnetic configuration. Here the difference between the two is the additional magnetic moment in the Cu^{2+}

5.3 Single Molecules: Magnetic Structure

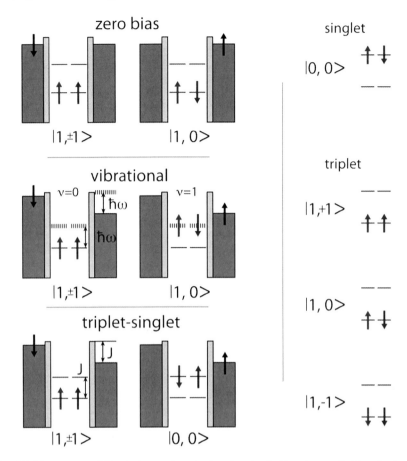

Fig. 5.8 Three types of Kondo interactions in the virtual spin flip picture: (i) The elastic zero bias Kondo spin flip scattering. The spin of the impurity is flipped by virtual excitations of electrons tunneling in and out of the impurity at zero energy cost (see Fig. 3.4 on p. 37). This event occurs in NiPc and CuPc molecules, and creates the characteristic Kondo resonance at E_F. (ii) Inelastic Kondo coupled to vibrational excitations. The coupling between the Kondo resonance and vibrational excitations of the molecule results in Kondo satellite peaks at the energy of the phonon in both molecules. (iii) Inelastic Kondo coupled to magnetic excitations, from the triplet ground state $|1\rangle$ to an singlet state $|0\rangle$. In CuPc this results in satellite peaks around the Kondo resonance. The *red arrows* indicate the ion spin (only for CuPc), the *blue* ones the ligand (for NiPc/CuPc). $|S, m_s\rangle$ represents the spin state of the system

ion, which coexist with the ligand spin in CuPc. We therefore assign these peaks observed in CuPc to intramolecular magnetic spin coupling, i.e., the coupling of the ϕ-spin to the d-spin forming a singlet ($S = 0$) and a triplet ($S = 1$) state. The transition between the two states is equal to the energy of the side peaks indicating an intramolecular exchange coupling of $J = 21 \pm 1$ meV. The fact that an intense Kondo peak is observed at zero bias indicates a magnetic ($S = 1$) ground state in the molecule, otherwise no Kondo screening would be possible. We thus conclude that the metal and ligand spin are aligned parallel to each other, which is supported by the gas-phase calculation of the magnetic moment of charged molecules presented in Sect. 5.3.2 on p. 89.

Underscreened Kondo Effect in CuPc

Although the magnetic configurations of CuPc ($S = 1$) and NiPc ($S = 1/2$) are different, the two molecules exhibit similar Kondo temperatures. This can be understood by considering that in CuPc only the ligand spin couples to conduction electrons of the substrate, providing a single screening channel. The effective decoupling of the $b_{1g}(d_{x^2-y^2})$ orbital, confined in the molecular plane, leads to an underscreened Kondo state, meaning that only part of the total spin $S = 1$ is screened and is reduced to $S = 1/2$ [24, 38, 45]. This is part of a two-stage process, where the presence of a strongly-coupled and a weakly-coupled screening channel implies the existence of two different Kondo energy scales [24, 45]. In the temperature range accessible to our experiment, only the $2e_g$ screening channel is effective, which explains why the measured Kondo temperature is similar for CuPc and NiPc. The fit of the intensity of the Kondo resonance using the analytic expressions $G_K(T)$ derived from renormalization group theory for the $S = 1/2$ and for the underscreened $S = 1$ Kondo effect [38] does not contradict this interpretation (Fig. 5.6b). Our measurements indicate that the residual spin of CuPc spin is not screened until temperatures much lower than 5 K.

Hence, the Kondo interaction in both molecules occurs only via the $2e_g$ spin, but in CuPc the latter is also coupled to the ion spin, as revealed by the spin excitations observed in the Kondo effect.

These excitations have been correctly reproduced by R. Korytar and N. Lorente by using a multiorbital noncrossing approximation model [44]. Here the coupling between ligand and ion spin is described by an intramolecular spin-coupling parameter I. A ferromagnetic coupling $I = 25$ meV describes the Kondo excitations of CuPc, reproducing the finite-bias features corresponding to inelastic Kondo replicas that originate from spin excitations. A zero coupling term $I = 0$ meV models the case of NiPc with only the ligand spin, as shown in the bottom graph in Fig. 5.7b.

Spatial Distribution of Inelastic Excitations

The coexistence of vibrational and triplet–singlet Kondo excitations together with the spatial resolution of STM allow us to map the spatial distribution of each type of inelastic Kondo excitation for CuPc. In Fig. 5.9 a series of spectra taken along a CuPc molecule, starting from a benzene ring up to the TM center are shown. The data show that the vibrational and triplet–singlet inelastic channels occupy mutually exclusive regions both in energy and space, with the former localized on the central part of the molecule and the latter on the external ring structure.

5.3 Single Molecules: Magnetic Structure

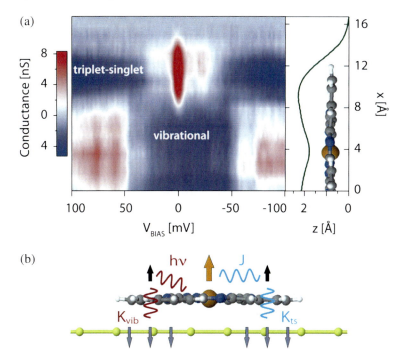

Fig. 5.9 a Series of dI/dV spectra acquired along a CuPc molecule revealing the different localization of the vibrational and triplet–singlet Kondo resonances. The position of each spectra is indicated by a topographic line scan on the right. **b** Schematic representation of the different coupling mechanisms behind each inelastic Kondo process

The localization of the vibrational coupling around the ion site can be understood on the grounds that all the observed mode involve distortions of Cu–N bonds (see Fig. 5.7). On the other hand, the intensity of the triplet–singlet excitations is proportional to the local spin/charge density as well as to the probability to tunnel to spin-polarized orbitals, the latter being much higher for the $2e_g$. The same reasoning applies to the zero bias Kondo peak itself, whose intensity mimics that of the $2e_g$ (see Fig. 5.5). Note that the two inelastic peaks exhibit a sharp spatial separation of around 1 Å.

Thus, as the STM tip has the role of a mobile electrode, our measurements show that the appearance of nonlinear resonances in the I/V curves of metal–organic complexes is related in a nontrivial way to the contact geometry with the metal leads and to the vibronic and magnetic degrees of freedom within the molecules. The atomic configuration of a molecule–metal junction determines not only the resonant transport through MOs, but also the type of inelastic processes that dominate transport at the Fermi region.

Relaxation and Universal Scaling of Nonequilibrium Kondo Excitations

The spin and current dynamics in small quantum objects coupled to external charge reservoirs are difficult to describe theoretically. This is especially true for systems where different many-body excitations are present [46]. Up until now this problem has been difficult to address experimentally. However we observe multiple relaxation channels for the CuPc ligand spin, related to nonequilibrium Kondo processes caused by vibrational excitations at energies E_{Kv1}, E_{Kv6} and a triplet–singlet transition at E_{Kts}. Hence CuPc provides the opportunity to compare spin relaxation via vibrational and magnetic cotunnelling events in the same molecule.

To do so, we deconvoluted the inelastic tunneling conductance of purely vibrational origin from the vibrational Kondo finite-bias intensity by fitting the dI/dV spectra of CuPc and NiPc using the sum of eight step functions and Lorentzian curves on each side of E_F (Fig. 5.7a). The triplet–singlet excitation features were modeled by two Lorentzian peaks as shown in Fig. 5.7b.

The full width at half maximum of the triplet–singlet peaks (Γ_{ts}) and vibrational resonances (Γ_v) is inversely related to the decoherence time of the cotunnelling processes schematized in Fig. 5.8. At temperature $T \geq T_K$, both Γ_{ts} and Γ_v are dominated by thermal broadening, similar to Γ_K, as shown in Fig. 5.6a. Deeper into the Kondo regime at $T < T_K$, however, we observe that Γ_{ts} deviates from Γ_K, saturating at 14 meV. This implies that the intrinsic timescales of triplet–singlet and zero-bias Kondo spin flips are different. These findings led us to investigate the scaling properties of the conductance as a function of T_K and V_b. Current theoretical models state that nonequilibrium Kondo physics is completely universal and determined by a single energy scale corresponding to T_K [47]. This universality, however, applies uniquely to the elastic Kondo resonance, hence there is no physical reason for Kondo processes that involve different type of inelastic excitations to obey the same decoherence rate as a function of energy.

Such a comparison can be carried out by renormalizing both Γ and V_b by the relevant Kondo energy scales by plotting Γ/Γ_K as a function of $V_b/k_B T_K$, in analogy with the theory of dc-biased quantum dots [47]. Figure 5.10 shows the renormalized Γ_{ts} and Γ_v values of CuPc and NiPc, obtained in this way. We find that Γ_v follows the exponential trend with V_b expected for the relaxation of nonequilibrium Kondo processes that are dominated by electron–hole pair excitations [24, 50]. However, Γ_{ts} of CuPc lies outside this trend, showing a larger normalized decoherence rate compared to vibrational excitations. By including data from literature experiments, with T_K differing by as much as two orders of magnitude, reporting singlet–triplet [38, 48], or vibrational [19, 49] Kondo features, we further support this conclusion.

Figure 5.10 shows that finite-bias resonances related to vibrational excitations fit a single exponential curve (red line) with a decay rate that is a factor 3 smaller compared with inelastic spin excitations (cyan line), reflecting the different decay channels present in each case. We conclude that the coherence of non-equilibrium Kondo cotunnelling events is universal only when restricted to the subspace of a given observable. Indeed, the faster decoherence, observed for the triplet–singlet channel, can be related to the higher coupling of the electron bath to spin excitations as com-

5.3 Single Molecules: Magnetic Structure

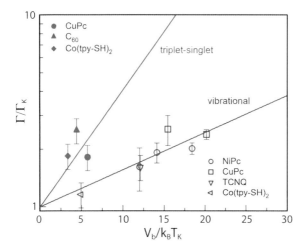

Fig. 5.10 Spin versus vibrational relaxation channels. Normalized width Γ/Γ_K of the inelastic Kondo resonances as a function of $V_b/k_B T_K$. Data for CuPc and NiPc are compared with that for other molecules for which either spin or vibrational Kondo excitations have been separately reported. Triplet–singlet (*cyan*) and vibrational (*red*) Kondo excitations separate into two groups. Exponential fits (*solid lines*) evidence the different contribution of each process to the decoherence rate. Data from C_{60} [48], Co(tpy-SH)$_2$ [38, 49] and TCNQ [19]

pared with phonon excitations [40]. A similar correlation between decoherence and inelastic channel might apply to other types of nonequilibrium Kondo phenomena, including those involving Cooper pairs or photon adsorption [51, 52].

5.3.2 DFT: Magnetic Structure

By calculating the saturation magnetic moments of different MePc, DFT provides complementary results to the analysis of the Kondo spectra reported above. Moreover, in addition to supporting the interpretation of our STS data, DFT allows us to construct a quantitative picture of the effect of the different charge transfer mechanisms on the molecular magnetic properties.

Table 5.1 summarizes the magnetic moments calculated for the neutral and anionic form of gas-phase MePc as well as for MePc adsorbed on Ag(100). Although the amount of charge transferred from the substrate to the molecules is similar in all cases, the total magnetic moment is reduced in FePc and CoPc, whereas it is increased in NiPc and CuPc with respect to the neutral gas-phase MePc. The calculations confirm that the magnetic moment of the TM ions in NiPc and CuPc is not perturbed upon adsorption, due to the small hybridization of the planar b_{1g} orbital with the substrate states. This is in agreement with previous results obtained by XMCD on CuPc/Ag(100) (see Fig. 5.11) and with the coexistence of TM and ligand spins in CuPc deduced from the Kondo spectra. We recall that the ligand spin is barely present

Table 5.1 Computed charge transfer ΔN (electrons) and total magnetic moment m (units of μ_B) in neutral and anionic gas-phase, and adsorbed molecules, calculated by GGA + vdW

	ΔN	m_{MPc}^{gas}	$m_{[MPc]^-}^{gas}$	m_{MPc}^{ads}
FePc	0.80	2.00	1.00	1.06
CoPc	0.99	1.00	0.00	0.63
NiPc	1.13	0.00	1.00	0.14
CuPc	0.81	1.00	2.00	1.32

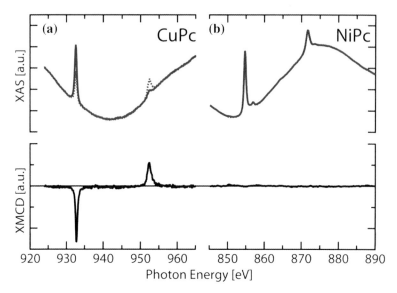

Fig. 5.11 X-ray absorption spectra (XAS) with circular polarized light of CuPc and NiPc monolayer films on Ag(100). The X-ray magnetic dichroism (XMCD) curve is derived from subtraction of XAS spectra with opposite circular polarization. **a** CuPc film showing an XMCD signal for the Cu edge, indicating that the Cu ion retains its moment. **b** For the NiPc film no XMCD signal is observed, meaning that the Ni is non-magnetic in the NiPc/Ag(100) configuration. Measurements taken at T = 8 K, normal incidence and 5 T applied magnetic field at the ID08 beamline of the ESRF. For more details see [53]

in the calculations with adsorbates due to the underestimation of correlation effects in DFT. For NiPc, we calculate a small magnetic moment of 0.14 μ_B, while for CuPc the moment is just slightly larger than the 1 μ_B corresponding to the unpaired spin in the b_{1g} orbital. Since the charge transfer is close to one electron, one way to estimate the magnitude of the ligand spin using DFT is to consider gas-phase anions as a model of an adsorbed MePc. Here the number of electrons belonging to the molecule is fixed and the calculations can be constrained to yield the minimum energy spin state. Figure 5.12 displays the spin density of neutral and anionic gas-phase molecules, where we clearly observe the additional ligand spin of NiPc and CuPc upon charge transfer. In CuPc, the spin density originating from the b_{1g} orbital is also observed, distributed over the Cu ion and N_p atoms.

5.3 Single Molecules: Magnetic Structure

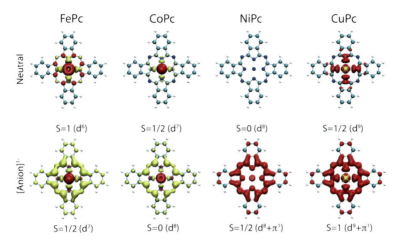

Fig. 5.12 Spin density of neutral (*top*) and anionic (*bottom*) gas-phase MePcs. *Red/yellow* indicate spin up/down. The iso-value for the spin density is 0.005

FePc and CoPc present a more complicate picture compared to the simple addition of one spin in NiPc and CuPc, partly because of the strong hybridization with the substrate discussed in Sect. 5.2.2. In both cases, the magnetic moment is reduced upon adsorption. The extra electron donated by the surface is not easily assigned to a single MO, inducing a reorganization of the charge and spin within the molecule.

The results obtained for gas-phase anions can be used to disentangle the different mechanisms responsible for the quenching of the magnetic moment. As shown in Fig. 5.13, the spin density of FePc follows the same distribution in the gas-phase anion

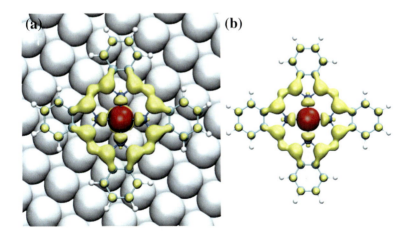

Fig. 5.13 Spin density distribution of FePc adsorbed on Ag(100) (**a**) and in the gas-phase anion (**b**). *Red/yellow* indicate spin up/down. The isovalue for the spin density is 0.005 in both cases

Table 5.2 Computed element-resolved magnetic moment m (units of μ_B) for FePc in the neutral and anionic gas-phase, and in the adsorbate, using GGA + U

	m_M	m_N	m_C	m_{tot}
FePc$_{gas}$	1.94	−0.13	0.19	2.00
[FePc]$^{1-}_{gas}$	2.04	−0.31	−0.73	1.00
FePc$_{Ag(100)}$	2.14	−0.26	−0.82	1.06

and in the adsorbed molecule. This resemblance is also confirmed by comparing their element-resolved magnetic moments in Table 5.2, which are very similar in both cases. Thus, regardless of the strong hybridization of adsorbed FePc, the main features of the spin moment distribution are again captured by the simple addition of one electron to the gas-phase system. The true nature of the ground state of FePc on Ag(100), however, is most likely more complex than suggested by the calculated spin density distribution, due to the interplay of several MO close to E_F and dynamic electron correlation effects that are not included in DFT. The computed electronic structure for FePc yields a S = 1/2 system when adsorbed on Ag(100), same as for the gas-phase anion. However, the adsorbate exhibits a more complex configuration due to the presence of multiple molecular orbitals at the Fermi level. This could result in either a mixed-valence system, given the Fermi level crossing and strong interaction of the a_{1g} orbital with the substrate, or a more complex Kondo behavior with a low-temperature Kondo phase originating in the many less-coupled orbitals near the Fermi energy. This complex behavior makes FePc/Ag(100) an interesting candidate to study very low temperature Kondo physics in a multi-orbital system.

CoPc represents another case of nontrivial charge transfer occurring upon adsorption. For example, the a_{1g} orbital is filled, but new empty d_π states appear above E_F, as shown in Fig. 5.2b, c. Moreover, the gas phase anion is not a good model for the adsorbed molecule. As shown in Table 5.1 and Fig. 5.12, the anion is in a S = 0 state, due to antiferromagnetic coupling between spins residing in different MOs, while in the adsorbed case we find a non-integer spin magnetic moment of $0.63\mu_B$. This non-integer spin, together with the absence of a Kondo interaction, could indicate that CoPc is in the mixed-valence regime, with charge fluctuating between the CoPc d and Ag states, similar to the results obtained for CoPc on Au(111) using X-ray absorption spectroscopy [4].

5.3.2.1 Intramolecular Spin Correlation

The TM ion-ligand spin coupling observed via the inelastic Kondo interaction in CuPc can be studied in gas-phase [CuPc]$^{1-}$. Here, the energy of ferromagnetic and antiferromagnetic spin configurations can be computed by fixing the total spin of the molecule. The ferromagnetic (triplet) alignment is favored over the antiferromagnetic (singlet) one. The triplet ground state agrees with the experimentally observed zero-bias Kondo resonance. From the energy difference between triplet and singlet states, evaluated using the parallel spin ($E_{\uparrow\uparrow}$) and antiparallel spin ($E_{\uparrow\downarrow}$) energies, and taking the correct singlet and triplet configurations to yield $J = 2(E_{\uparrow\downarrow} - E_{\uparrow\uparrow})$, the exchange coupling constant $J = 38$ meV for the anion can be extracted. The discrepancy with the experimental value $J = 21$ meV for CuPc/Ag(100) can be explained by surface screening, which is expected to significantly reduce exchange correlation effects involving hybridized orbitals.

The distribution of spin density in [FePc]$^{1-}$ and [CoPc]$^{1-}$ shown in Fig. 5.12 reveals a different type of intramolecular spin correlation. Both neutral and anionic molecules exhibit antiferromagnetically aligned TM ion and ligand spins, indicating that the contrast is not related to new spins induced by the substrate, but is intrinsic to the molecule. It is derived from the original TM ion spin, which extends to nearby C and N atoms upon the formation of MOs with mixed d- and π-character. This intramolecular antiferromagnetic spin coupling is a result of the large exchange splitting of the d-levels of the Fe and Co ions. It induces spin-dependent mixing of the d and ligand states, leading to the formation of spin-polarized MOs with different spatial distribution. The site and energy dependent spin contrast recently measured for CoPc deposited on Fe(110), with a total molecular moment of zero, could be explained by a similar mechanism [13].

5.4 From Clusters to Monolayer

As we have seen, the electronic and magnetic characteristics of single MePc molecules are strongly influenced by their interaction with the surface. Intermolecular interactions can, however, play a role as well. Previous experiments found a decoupling caused by increased molecular coverage: For instance an STM study of a FePc monolayer on Ag(111) showed that intramolecular interaction leads to sharper single resonances in STS [16]. Photoemission experiments for CuPc on Ag(111) also came to this conclusion [54].

Here we investigate this in more detail by increasing the number of neighboring molecules step-by-step, and their effect on the evolution of the electronic structure. CoPc and CuPc molecules serve as model systems for the main hybridization channels through ion's d states and the ligand orbitals respectively. We will see that the interaction with the surface through the more delocalized ligand states, as is the case for CuPc is strongly influenced by intermolecular forces, leading to complex changes in the electronic structure. In contrast, for CoPc, the interaction through hybridized d states is unaffected.

5.4.1 Small Clusters of CuPc

To systematically investigate the influence of neighboring molecules on the electronic and magnetic structure, we studied small CuPc clusters. Using the lateral

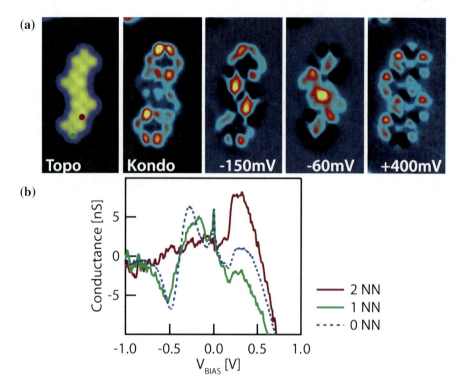

Fig. 5.14 a STM topography of a 4 × 1 CuPc cluster, and dI/dV maps of STS resonances at the indicated energies. The Kondo map is measured using the d^2I/dV^2 signal at −6 mV. Note that the Kondo resonance can be observed for all four molecules, albeit weaker for the center ones. b dI/dV spectra taken on the outer benzene of a CuPc with 1 NN and on the bonding edge of a CuPc with 2 NN compared to a single CuPc with no NN (3.0 nA, −1.0 V). The positions of the dI/dV spectra are indicated in the topography. The spectra haven been background subtracted

manipulation capabilities of our STM, we created several cluster geometries with different nearest neighbor configurations, in a controlled and reproducible way (see Fig. 2.9 on p. 23). We ensured that the molecules remained unperturbed and in the same adsorption configuration after manipulation by comparing topographic images and conductance spectra before and after their displacement.

Figure 5.14a shows a linear cluster consisting of four CuPc. The two molecules at the end have 1 nearest neighbor (NN), while the two central ones have 2 NN. We investigate the impact of NN on the electronic structure by dI/dV spectra maps.

The spectra for molecules with 1 NN is very similar to that of the single molecules; the occupied and unoccupied MO around E_F are still present as well as the Kondo resonance. We observe a small upshift of the charged peak, previously attributed to the $2e_g$ state (see Fig. 5.14b). The spatial distributions of the states found above and below E_F differ however. The two dI/dV maps taken at −150 mV and +400 mV are orthogonal to each other, indicating that they not caused by the same state anymore. The twofold orbital-degeneracy of the $2e_g$ state is hence broken by the interaction

5.4 From Clusters to Monolayer

with the neighboring molecule. The two new states are labeled as α_{2eg} (the single occupied orbital) and β_{2eg} (the unoccupied orbital). The electron that is transferred from the substrate is located in the α_{2eg}, and is now the responsible for the Kondo interaction. This can be seen by the similarity with the Kondo intensity distribution: Both dI/dV maps have a two-fold symmetry instead of the four-fold symmetry of the $2e_g$. The unoccupied coulomb pair is not visible, probably due to the overlapping with the more intense β_{2eg}. The orbital rearrangement coincides with a lowering of the Kondo temperature (see Fig. 5.16).

The electronic structure of the two central molecules with 2 NN shows no further changes of the unoccupied β_{2eg} state. The occupied α_{2eg} orbital however is not observable in the spectra, nor is its symmetry found in the dI/dV maps. Instead, a new feature appears at ~ -60 mV with a different intensity distribution, mainly centered between the molecules. On the other hand, the Kondo resonance is visible and follows the same symmetry as for molecules with 1 NN albeit with a weaker intensity, and a further downshifted T_K (Fig. 5.16). Its presence and symmetry indicate that the α_{2eg} orbital is still singly occupied for molecules with 2 NN. In the STS measurements it is most likely concealed by the more intense new state at -60 mV, whose origin remains unclear.

To investigate molecules with more NN, we switch to a different cluster geometry. Figure 5.15 shows a 3×3 square cluster created from a larger self-assembled structure through lateral manipulation (see p. 23). The corner molecules have 2 NN, the central molecules on the sides have 3 NN, and the center one has 4 NN. The dI/dV maps of the corner molecules show a behavior similar to that found for molecules with 2 NN in the 4×1 cluster, where the unoccupied region changed. Molecule with 3 NN and 4 NN, however, show changes occurring to both the occupied and unoccupied states. The state at -80 mV is pronounced and has most of its intensity between the molecules in the cluster. The state above E_F is shifted upward from $+300$ mV to $+650$ mV for 3 NN and to $+950$ mV for 4 NN. Its intensity distribution changes with the bonding environment. This upshift points towards a decoupling from the substrate, which coincides with the complete disappearance of the Kondo resonance found for 3 NN and above (see Fig. 5.16).

Note that this change is reversible. By removing one of the corner molecules of the cluster two molecules change form 3 NN to 2 NN, and show the Kondo resonance and the unoccupied peak at $+300$ mV both characteristic of molecules with 2 NN (Fig. 5.15a). The fact that the Kondo interaction can be "turned on" proves that the molecules are all unperturbed, and that its gradual quenching is exclusively related to the number of NN.

Previous studies have shown that the Kondo temperature of metal–organic complexes can be manipulated by locally modifying the ligand field acting on the transition-metal ions [38, 55] or by screening the surface electron density around molecular adsorbates by other molecules [56]. In our system, we observe a gradual lowering of the Kondo temperature with increasing NN, which is related to an evolution of the molecular electronic structure. It is difficult, however difficult to specifically connect the disappearance of the Kondo resonance with the emergence of a new state below E_F, or the decoupling of the molecule suggested by the upward

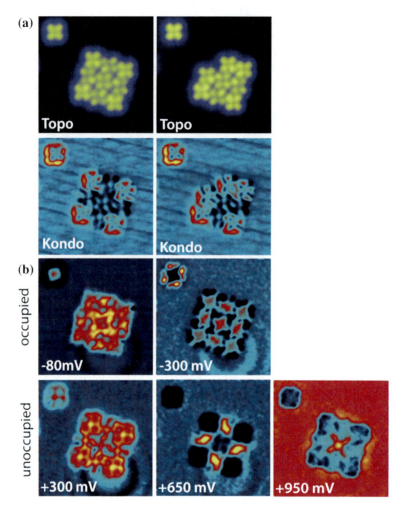

Fig. 5.15 a A 3 × 3 cluster of CuPc before (*left*) and after (*right*) removing a corner molecule by manipulation with the tip: Topography and d^2I/dV^2 maps at −6 mV showing the Kondo intensity distribution (9.7 nm × 9.7 nm). Note that the Kondo resonance can be observed only for molecules with a number of lateral bonds smaller than three. **b** dI/dV conductance maps of the 3x3 cluster recorded at the energies of different molecular orbitals

shift of the β_{2eg}. In any case, both effects indicate a strong distortion of the original molecular states.

5.4 From Clusters to Monolayer

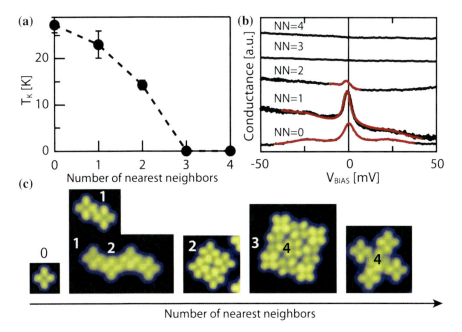

Fig. 5.16 **a** Evolution of the Kondo temperature with the number of intermolecular bonds, obtained from Fano function fits to the dI/dV shown in **b** acquired at 5 K at the benzene of molecules with different number of nearest neighbors and in different configurations. **c** Examples of CuPc clusters used in the study of the dependence of the Kondo interaction on the number of intermolecular bonds, indicated with numbers

5.4.2 Monolayers of CuPc and CoPc

CuPc Monolayer

To complete the investigation of the influence of intermolecular forces on the electronic structure we investigated a close-packed monolayer of CuPc. Since the molecules in the monolayer are also in a 4 NN configuration, the STS curves are identical.

The state at −80 mV has its maximum intensity between the molecules of the monolayer, as can be seen in the dI/dV maps and spectra shown in Fig. 5.17. For the unoccupied states, we observe a progression of sharp resonances, depending on where the spectra was taken exactly: on the ion (+650 mV), between the molecules (+750 mV), and on the benzene (+790 mV). These peak energies show an extreme dependence on the tip height during the measurement. Kröger et al. have shown that, even within the tunneling regime of an STM, the electric field produced by the tip can affect the energy position of the surface state, analogue to the stark effect [57]. Molecular orbitals sufficiently decoupled from the metal substrate and with the right alignment in the electric field, should also be shifted by this tip induced Stark shift [58]. Indeed, the plot of the resistance against the peak position (Fig. 5.18a) shows

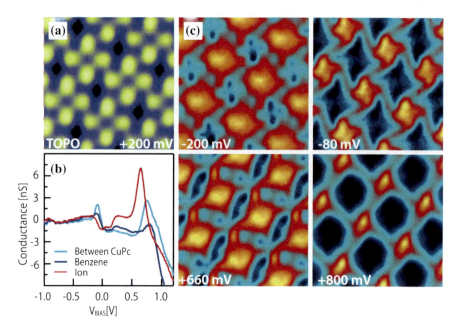

Fig. 5.17 Electronic structure of 1 monolayer CuPc/Ag(100). **a** STM Topography (0.5 nA, +0.2 V, 4.2 nm × 4.2 nm). **b** dI/dV spectra at different points of the molecule within the monolayer (3.0 nA, −1.0 V). **c** dI/dV maps of the same region, showing the molecular resonances near the Fermi level. The spectra have been background subtracted for clarity

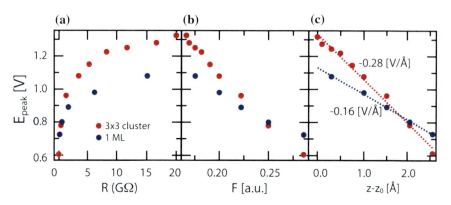

Fig. 5.18 Stark-shift of the β_{2eg} level: **a** Peak position versus tunneling resistivity. **b** Peak position versus electric field, calculated following Ref. [57]. **c** Linear relationship between tip height z and the peak position. Positive z values correspond to a smaller tip-molecule distance. The tunneling gap was set at 0.015 nA, +0.4 V

the characteristics of a linear Stark effect. This becomes even clearer when plotting the shift versus the electric field, using the method from Ref. [57]. The peak position versus tip height reveals a linear dependence −0.16 V/Å. The molecule with 4 NN in the 3 × 3 cluster (Fig. 5.15) also shows this behavior, however with a higher shift of −0.28 V/Å, this variance is most likely caused by the different fields created by different tips.

The origin of the position dependence of the unoccupied features in the dI/dV spectra cannot be related to the observed Stark effect. The topography indicates that the tip moves towards the sample between the molecules, which for the Stark effect would indicate an upshift in energy. However, we observe the reverse effect: the feature between the molecules lies lower than that on the benzene. A possible explanation might be related to a band dispersion due to hybridization with the surface state, as observed for PTCDA islands on Ag(111) [59, 60].

CoPc Monolayer

In contrast to the pronounced changes occurring in the CuPc monolayer, the spectra for CoPc show no difference between the single CoPc and the monolayer (see Fig. 5.19). Hence, CoPc does not decouple from Ag(100) by intermolecular interactions.

The coupling to the substrate in CoPc occurs mainly through the TM's d states, while for CuPc the ligand states are the main interaction channel (see Table 4.1 on p. 53). The difference in behavior for CuPc and CoPc could be related to (i) the degree of the coupling which appears to be stronger for CoPc with direct TM-substrate interaction, and (ii) the fact that intermolecular interactions directly involve the ligand orbitals, and hence could possibly weaken their coupling to the substrate. That would only affect the electronic structure and/or adsorption configuration of CuPc, where the charge state of the ligand orbital depends on the coupling to the substrate.

The coupling strength, however, also depends on the substrate. Reference [16] reports a decoupling for a FePc monolayer on Ag(111) driven by intermolecular forces. FePc interacts with the substrate in roughly the same way as CoPc, through strongly hybridized d and ligand states. The weaker interaction of absorbates with Ag(111) seem to allow intermolecular forces to decrease molecule substrate coupling even for molecule of the FePc/CoPc type [16].

5.5 CuPc Multilayer

We have seen so far that MePcs absorbed on a surface are heavily influenced by the interaction with the substrate. Molecular states are broadened and shifted by charge transfer and hybridization with the substrate. Even the electronic structure of a CuPc monolayer, where lateral molecular interactions become important, was

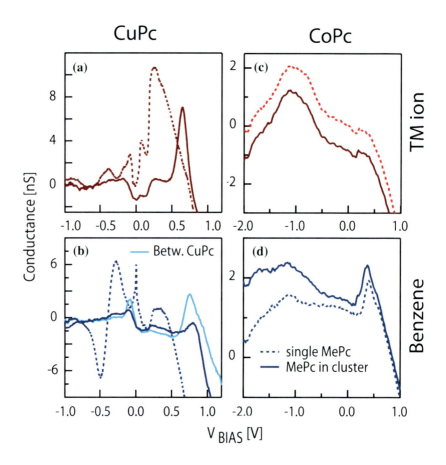

Fig. 5.19 dI/dV spectra taken on CuPc and CoPc inside a cluster (*solid line*), and single molecules (*dotted line*) on the ion (**a** and **c**) on the benzene ring (**b** and **d**). For CuPc an additional spectrum was taken between neighboring molecules (*light blue*). The set points were 3.0 nA, −1.0 V for CuPc, and 2.0 nA, −2.0 V for CoPc, all spectra are background subtracted

still dominated by substrate-molecule interaction. In this section we will study the evolution of the electronic structure of CuPc stacked up to 5 molecular layers as it changes towards the semiconducting behavior of bulk CuPc. The influence of the substrate is gradually weakened by the lower lying molecular layers, and the molecules become more and more decoupled from the surface. We observe effects related to this decoupling such as sharpening of the molecular resonances [61, 62] and an increase in electronic lifetime that leads to a coupling to molecular phonons [63]. For thick layers, a second tunneling barrier is created in addition to the vacuum region between the STM tip and the topmost molecule, leading to interesting phenomena caused by this double tunnel junction geometry, such as bipolar tunneling [64] and negative differential resistance (NDR) [65].

5.5 CuPc Multilayer 101

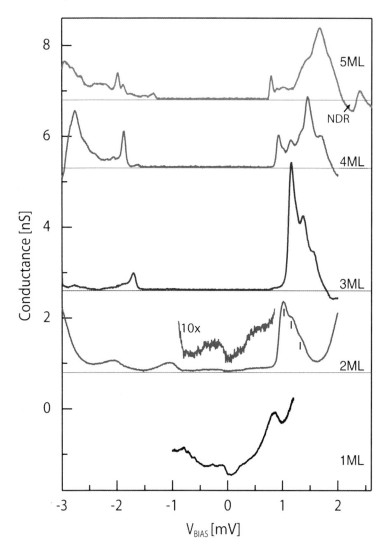

Fig. 5.20 Evolution of the electronic structure of multilayer CuPc/Ag(100) as a function of film thickness (1–5 ML). The gap increases, and vibronic features (ticks) and regions with NDR appear with increasing thickness, all signatures of an effective decoupling from the metallic substrate. The zero line for the spectra is indicated. All spectra are taken on a benzene ring, the vacuum gap was set to 0.2 nA, −3.0 V except for the 1 ML spectrum, for which 3.0 nA, −1.0 V was used

5.5.1 Molecules on Higher Layers: Spectroscopy

We performed STS measurements on CuPc molecules in different layers. The evolution of the spectra can be seen in Fig. 5.20. The dI/dV taken on the second layer changes drastically compared to the monolayer spectrum. A gap opens around E_F, as peaks shift further apart. The onset of these features lie at -0.88 V and $+0.86$ V respectively, implying a gap of around $+1.70$ V. Moreover, the spatial intensity distributions of the peaks closely follow the HOMO and LUMO states observed for CuPc on Ag(100) (see Fig. 5.3). The conductance around E_F, however, is still not completely zero, and two weak and broad peaks are found at -0.28 V and $+0.43$ V. Furthermore, we observe that the positive bias peak at $+1.00$ V is made up of three overlapping peaks. This multi-peak structure is caused by vibronic excitations, which have been observed for various molecules on insulating substrates [63, 64, 66], and will be discussed in more detail in Sect. 5.5.1. The appearance of these effects, the gap opening and the vibronic coupling are clear signs that the molecules decouple from the surface [16]. However, the fact that two peaks lie in the gap indicates that this decoupling is not yet complete as these two peaks could be remnants of the interaction with the substrate or an interface state. The weaker coupling to the substrate is such that it leaves the main characteristics of gas-phase CuPc unperturbed, in contrast to what we have seen for single molecules and the monolayer (Fig. 5.21). The magnitude of the gap (1.70 V) is very close to the HOMO–LUMO gap measured for thin CuPc films by PES/IPES (1.50 V [67] and 1.70 V [68]).

For molecules on the 3rd, 4th, and 5th layer the spectra show additional changes. The peaks become sharper, and the multiple peak at positive bias shows even more features. The gap between states significantly increases to roughly 2.50 V, while the STS signal inside the gap is zero. Lastly at the far positive end of the spectra we measure negative differential conductance (NDR). The zero conductance gap, the sharply defined vibronic features and the NDR are all features that have been observed for molecules adsorbed on insulating layers [63, 64, 69]. Hence the two underlying organic layers effectively decouple the molecules on the 3rd layer and above from the substrate.

There are now two tunneling barriers, the vacuum between tip and molecules, and the organic layers underneath. This situation makes it difficult to assign spectral features to MO and the wider gap has to be considered on these grounds.

A certain part of the bias voltage is dropped over the insulating layer, while the rest applies to the vacuum gap. This voltage drop makes both tunneling barriers voltage dependent, resulting in bipolar tunneling [69]. This means it is possible to tunnel through the same MO orbital at different bias voltages with opposite signs. Either the LUMO or the HOMO will be observed above and below E_F.[1] This depends on the ratio between the two barriers [63], and the distance of the HOMO/LUMO level to E_F [62]. For CuPc on thin NiAl oxide films, even the molecular adsorption conformation was found to influence whether the HOMO or LUMO state was accessed by STS [64]. Unfortunately no values for the tip height z were measured during the experiment, so that the barriers cannot be quantified, making it difficult to determine if the observed states are related to the LUMO or the HOMO. Moreover, the energy positions at which molecular resonances are observed, depend on the bipolar tunneling. Hence

---

[1] In special conditions it is also possible to the see the LUMO above and the HOMO below E_F, similar to single tunnel barrier system [62].

5.5 CuPc Multilayer

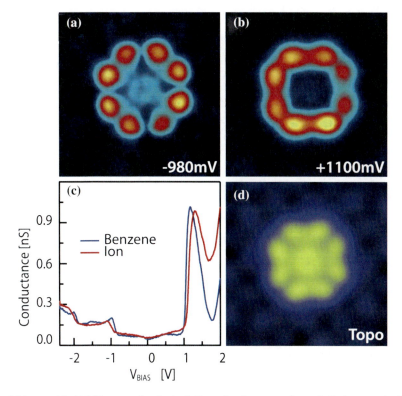

Fig. 5.21 a and **b** dI/dV maps of a single CuPc molecule on top of one CuPc layer on Ag(100) (5.3 nm × 5.3 nm). The HOMO and LUMO distributions are clearly visible for −980 mV and +1100 mV. **c** dI/dV spectra taken on a single CuPc on the 2nd layer (0.2 nA, −3.0 V). **d** A topographic image obtained during the acquisition of the dI/dV maps (0.01 nA, +1.2 V)

the increase of the gap above 2 ML is probably related to this phenomena, possibly in combination with screening effects due to the polarizable neighborhood of the molecules [16, 70, 71].

The NDR, observed in Fig. 5.20 above +2 V is another effect caused by voltage dependent tunneling barriers in combination with sharp resonances and the absence of broad metallic states [65, 72]. Its presence further confirms the that molecules in the 3rd layer are already decoupled from substrate.

The fact that the spectra for the 3rd, 4th, and 5th are relatively similar, except for intensity changes in the multi-featured peak, indicates that the decoupling from the substrate is completed on the 3rd layer. As the tunneling rate, i.e., the current, was kept constant between the spectra, effects due to the changing barriers thicknesses (higher molecular layers), were counteracted by adjusting the tip height z accordingly.

Vibronic Progressions

The positive bias peak around +1.6 V is a multi-featured structure for the spectra taken on thicker layers. It was shown by W. Ho and co-workers [63, 64, 66] that a molecular orbital on an insulating layer can couple to vibrational excitation states, which will then be visible as *equidistant* peaks overlaying the actual molecular orbital. In these works the insulating layer was a thin NiAl oxide film. In our system already one molecular layer leads to a sufficient decoupling for vibronic excitations to be measured. With increasing layer thickness the quality of the decoupling increases and sharper and more vibronic peaks appear.

In multilayer CuPc all of the positive bias peaks for all layer thickness can be fitted in this manner. Figure 5.22 shows the results of multi peak fits for the positive bias peak, with multiple Lorentzian revealing roughly equidistant peaks. We hence attribute the origin of these peaks to vibronic coupling to a MO. For each layer a slightly different energy spacing is found: 153 mV for the 2 ML, 155 mV for the 3 ML, 130 mV for the 4 ML and 172 mV for the 5 ML. Literature values for C–C or C–N stretching modes lie exactly within this energy range (150–200 meV) [73]. Furthermore PES measurements for thin films of CuPc on HOGP [73, 74], and in the gas-phase [75] show a vibronic coupling of the CuPc's HOMO to a vibronic excitation with 150 meV. The energy difference between the layers is most likely due to the fact that these energy values are also affected by the double tunnel junction and might vary by as much as 5–20 % depending on the exact geometry of the tunnel barrier [66].

We note that on the third layer part of the molecules lie flat, while others are already in the tilted geometry observed for the next layers. Spectroscopy taken on these two types of molecules on the 3rd layer allows us to study the effect of the tilt on the electronic structure (see Fig. 5.23). Both spectra are very similar, however the onset of the positive bias peak has an enhanced feature in the tilted configuration. The fit of the complete multi-peak reveals that this peak forms part of the vibronic progression. In the panel for the 3 ML in Fig. 5.22 it can be seen that the peak positions for flat (red dots) and tilted (all other) molecules are offset with respect to each other. The energy spacing for both geometries, however, was found to be roughly 150 mV.

The 4th layer shows another example of different intensities of the vibrational peaks. Here the benzene rings of opposite axis are in different configurations, due to the tilted adsorption, leading to changes in the relative intensity of the vibronic peak changes (see Fig. 5.23c, d).

These two observations are related to the difference in molecular configuration, it is however not clear, whether it caused by changes in the Frank–Condon factors that determine the intensities of vibronic transitions, or by the tunneling barrier and hence the peak position, which was also found to be affected by the molecular configuration [64].

5.5 CuPc Multilayer

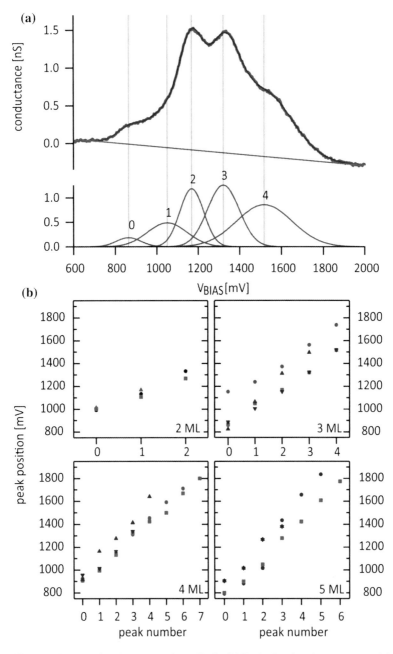

Fig. 5.22 **a** Fit of the multi peak structure above E_F for 3 ML, the fitted peaks are presented. **b** Peak position obtained from similar fits for the different coverages at different positions on the molecule, a roughly equidistant distribution is visible, pointing towards a vibronic origin of the multipeak structure. For the 3 ML the *red dots* correspond to the flat configuration, all other symbols to tilted molecules

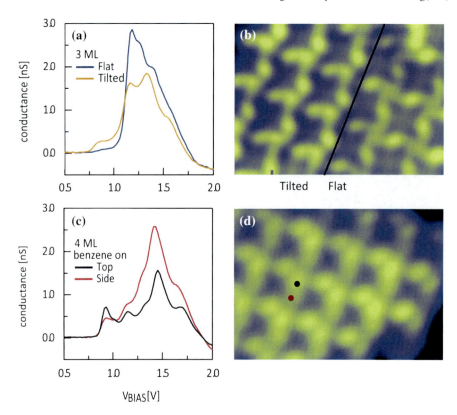

Fig. 5.23 **a** dI/dV taken on CuPc in the 3rd layer in the tilted and the flat geometry (0.2 nA, −3.0 V). **b** STM topography showing the region with tilted and flat CuPc in the 3rd layer (7 nm × 5 nm, 0.01 nA, 1.5 V). **c** dI/dV spectra taken on CuPc in the 4th layer (0.2 nA, −3.0 V). The intensity of different vibronic peaks shows a strong dependence on the site of the spectra. **d** STM topography of CuPc in the 4th layer indicating the points where the spectra in **c** were taken (7 nm × 5 nm, 0.01 nA, 1.4 V)

5.6 Summary

Single Molecules

We studied the electronic and magnetic structure of four MePcs (Me = Fe, Co, Ni, Cu) adsorbed on Ag(100), investigating systematically the influence of the central metal ion on charge transfer, hybridization and magnetic properties.

Charge transfer and hybridization: The electronic structure of adsorbed MePc is modified mainly due to charge transfer from the substrate. Although all molecules receive approximately one electron, charge is transferred to the ligand $2e_g$ orbital in NiPc and CuPc, and to multiple MOs in FePc and CoPc, inducing internal charge reorganization. This difference is related to both energy level alignment and Coulomb

5.6 Summary

repulsion of ligand and metal orbitals. In addition the degree of hybridization of the MePc's d states is based on their spatial distribution, i.e., in-plane or out-of-plane. FePc and CoPc exhibit strongly hybridized $d_\perp$ states near E_F, whereas in NiPc and CuPc these states are shifted down in energy, and instead the $d_\parallel$ states that are unperturbed upon adsorption appear at this energy range.

Reduction of the magnetic moment in FePc and CoPc: Both charge transfer and hybridization tend to reduce the magnetic moment of FePc and CoPc. The computed electronic structure of FePc indicates a $S = 1/2$ system when adsorbed on Ag(100). However the strong hybridization of MOs near the Fermi energy could indicate either a mixed-valence system, given the strong interaction of the a_{1g} orbital with the substrate, or a more complex Kondo behavior with a low-temperature Kondo phase originating in the many less-coupled orbitals near the Fermi energy. The non-integer moment obtained for CoPc and the existence of a substantial molecular DOS at the Fermi energy, together with the absence of a Kondo resonance in this molecule, suggest a mixed-valence configuration.

Additional magnetic moment in NiPc and CuPc: Opposite to the moment reduction in FePc and CoPc, the spin multiplicity is actually enhanced by charge transfer in NiPc and CuPc, due to the introduction of a ligand spin in the $2e_g$ orbital. The magnetic moment of the ion remains unperturbed. This transforms NiPc into a paramagnetic molecule and induces a triplet ground state in CuPc, where the ligand spin is exchange-coupled to the ion magnetic moment.

Ligand Kondo effect in NiPc and CuPc: The interaction between the $2e_g$ ligand spin of NiPc and CuPc and the substrate gives rise to a Kondo interaction, which induces a prominent zero bias resonance delocalized over the organic macrocycle of the molecules. In CuPc, the total magnetic moment is underscreened as the Kondo energy scale of the TM spin is orders of magnitude smaller than the one of the $2e_g$ ligand spin.

Inelastic excitations: The coherent coupling between Kondo spin flip and vibrational excitations induces inelastic Kondo resonances in both NiPc and CuPc. Further in CuPc intramolecular exchange interaction between the ion and ligand spin is revealed through the coupling of Kondo and triplet–singlet excitations. The coexistence of different non-equilibrium Kondo processes related to vibrational and spin transitions in CuPc opens up the possibility to study the timescale and spatial localization of multiple spin excitation and relaxation channels within a single molecule.

In general, we show that charge transfer and hybridization of MOs orbitals have profound effects on the electronic configuration, magnetic moment, and transport properties of metal–organic complexes adsorbed on a metallic substrate. Such effects in MePc do not only depend on the electronic configuration of the TM ions. Rather, the energy position, symmetry, and spin polarization of the pristine ligand and ion MOs has to be considered within a comprehensive picture of the charge transfer process. Importantly, the magnetic moment of adsorbed MePc can be either reduced or enhanced depending on the relative energy of the $d_\perp$ and π-levels, and their degree of hybridization with the substrate. Further, the possibility of inducing additional ligand spins upon adsorption may be relevant to establish magnetic coupling in extended molecular structures assembled on metal surfaces.

Clusters and Monolayer

Creating artificial clusters, we have studied in a bond by bond manner the influence of neighboring molecules on the electronic structure. For CuPc on Ag(100) we observe an evolution of the electronic structure that is more complex than a simple decoupling, highlighting effects due to lateral neighbors.

Evolution of the electronic structure: For CuPc already the interaction with 1 NN molecule leads to a break of the $2e_g$ degeneracy into two orthogonal states, the partially occupied α_{2eg} and the unoccupied β_{2eg}. The ligand Kondo resonance is still present, caused by the single electron in the α_{2eg}. Starting with 2 NN a new state at ~ -80 mV, which is mainly located between the molecules, is formed. The origin of this state remains unclear. CuPc molecules with 3 or 4 NN become more decoupled from the substrate, as seen by an upshift of the β_{2eg}, and a Stark shift of the same orbital caused by the electric field of the tip. The β_{2eg} state shows a position dependence of its energy position, which might indicate a band formation, probably mediated by hybridization with a surface state.

Evolution of the Kondo resonance: The Kondo temperature and intensity are gradually lowered for each additional NN, until the resonance completely disappears for CuPc with 3 or more NN. Its quenching correlates with the evolution of the electronic structure. Although its coincidence with the upshift of the β_{2eg} orbital is evident, the evolution of the spin of the unpaired electron remains unclear.

Rigid substrate-coupling for CoPc: We do not see any difference between single molecule and a monolayer for CoPc, due to the fact that the interacting states are localized at the metal ion and thus more protected from intermolecular interactions, and probably also due to the stronger molecule–substrate interaction than in CuPc.

We have shown that lateral molecular interactions can lead to a change in coupling between molecule and substrate. The effect of lateral interaction depends on the coupling strength and channel (either the TM's d states or the ligand), and ranges from a complex evolution of states for CuPc to no effect for CoPc.

Multilayer

To conclude we explored the electronic structure of up to five molecular layers of CuPc on Ag(100). We observe layer dependent decoupling from the substrate:

Partial decoupling for 2 ML: Molecules on the second layer exhibit signs of a decoupling from the underlying substrate. A band gap opens between the HOMO and the LUMO, showing a value similar to that reported in literature (1.7 V). The decoupling is however not yet complete, indicated by weak features within the gap.

Complete decoupling from 3 ML: From the 3rd layer on, the molecules become decoupled form the substrate, as evidenced by sharp spectral features, the appearance of a double barrier junction with voltage depended tunneling barriers, leading to NDR and strong shifts is energy, possibly due to bipolar tunneling.

Vibronic progressions: For all layers (2nd, 3rd, 4th, 5th) vibronic excitation peaks are observed. This is more clear for the positive peaks. The observed

5.6 Summary

vibrational energies fit quite well to values reported in literature. Additionally, we observe effects of the molecular configuration modulating the intensity distribution of the vibrational peaks of the molecule.

In conclusion the vertical decoupling becomes efficient from the 3rd molecular layer on, while the 2nd layer appears to be in an intermediate state, sufficiently decoupled to show the gas-phase HOMO–LUMO gap, and vibronic coupling, but not completely to provide a second tunneling barrier.

References

1. P. Gambardella, S. Stepanow, A. Dmitriev, J. Honolka, F.M.F. de Groot, M. Lingenfelder, S.S. Gupta, D.D. Sarma, P. Bencok, S. Stanescu, S. Clair, S. Pons, N. Lin, A.P. Seitsonen, H. Brune, J.V. Barth, K. Kern, Supramolecular control of the magnetic anisotropy in two-dimensional high-spin Fe arrays at a metal interface. Nat. Mater. **8**(3), 189–193 (2009). doi:10.1038/nmat2376
2. Z. Hu, B. Li, A. Zhao, J. Yang, J.G. Hou, Electronic and magnetic properties of metal phthalocyanines on Au(111) surface: a first-principles study. J. Phys. Chem. C **112**(35), 13650–13655 (2008). doi:10.1021/jp8043048
3. X. Chen, M. Alouani, Effect of metallic surfaces on the electronic structure, magnetism, and transport properties of Co-phthalocyanine molecules. Phys. Rev. B **82**(9), 094443 (2010). doi:10.1103/PhysRevB.82.094443
4. S. Stepanow, P.S. Miedema, A. Mugarza, G. Ceballos, P. Moras, J.C. Cezar, C. Carbone, F.M.F. de Groot, P. Gambardella, Mixed-valence behavior and strong correlation effects of metal phthalocyanines adsorbed on metals. Phys. Rev. B **83**(22), 220401 (2011). doi:10.1103/PhysRevB.83.220401
5. Y. Fu, S. Ji, X. Chen, X. Ma, R. Wu, C. Wang, W. Duan, X. Qiu, B. Sun, P. Zhang, J. Jia, Q. Xue, Manipulating the Kondo resonance through quantum size effects. Phys. Rev. Lett. **99**(25), 256601 (2007). doi:10.1103/PhysRevLett.99.256601
6. L. Gao, W. Ji, Y.B. Hu, Z.H. Cheng, Z.T. Deng, Q. Liu, N. Jiang, X. Lin, W. Guo, S.X. Du, W.A. Hofer, X.C. Xie, H. Gao, Site-specific Kondo effect at ambient temperatures in iron-based molecules. Phys. Rev. Lett. **99**(10), 106402 (2007). doi:10.1103/PhysRevLett.99.106402
7. N. Tsukahara, K.-i. Noto, M. Ohara, S. Shiraki, N. Takagi, Y. Takata, J. Miyawaki, M. Taguchi, A. Chainani, S. Shin, M. Kawai, Adsorption-induced switching of magnetic anisotropy in a single iron(II) phthalocyanine molecule on an oxidized Cu(110) surface. Phys. Rev. Lett. **102**(16), 167203 (2009). doi:10.1103/PhysRevLett.102.167203
8. J.D. Baran, J.A. Larsson, A.A. Cafolla, K. Schulte, V.R. Dhanak, Theoretical and experimental comparison of SnPc, PbPc, and CoPc adsorption on Ag(111). Phys. Rev. B **81**(7), 075413 (2010). doi:10.1103/PhysRevB.81.075413
9. A. Zhao, Q. Li, L. Chen, H. Xiang, W. Wang, S. Pan, B. Wang, X. Xiao, J. Yang, J.G. Hou, Q. Zhu, Controlling the Kondo effect of an adsorbed magnetic ion through its chemical bonding. Science **309**(September), 1542–1544 (2005a)
10. Z. Li, B. Li, J. Yang, J.G. Hou, Single-molecule chemistry of metal phthalocyanine on noble metal surfaces. Acc. Chem. Res. **43**(7), 954–962 (2010). doi:10.1021/ar9001558
11. C. Iacovita, M. Rastei, B. Heinrich, T. Brumme, J. Kortus, L. Limot, J. Bucher, Visualizing the spin of individual cobalt-phthalocyanine molecules. Phys. Rev. Lett. **101**(11), 116602 (2008). doi:10.1103/PhysRevLett.101.116602
12. B.W. Heinrich, C. Iacovita, T. Brumme, D. Choi, L. Limot, M.V. Rastei, W.A. Hofer, J. Kortus, J. Bucher, Direct observation of the tunneling channels of a chemisorbed molecule. J. Phys. Chem. Lett. **1**(10), 1517–1523 (2010). doi:10.1021/jz100346a

13. J. Brede, N. Atodiresei, S. Kuck, P. Lazicacute, V. Caciuc, Y. Morikawa, G. Hoffmann, S. Blügel, R. Wiesendanger, Spin- and energy-dependent tunneling through a single molecule with intramolecular spatial resolution. Phys. Rev. Lett. **105**(4), 047204 (2010). doi:10.1103/PhysRevLett.105.047204
14. P. Gargiani, M. Angelucci, C. Mariani, M.G. Betti, Metal-phthalocyanine chains on the Au(110) surface: interaction states versus d-metal states occupancy. Phys. Rev. B **81**(8), 085412 (2010). doi:10.1103/PhysRevB.81.085412
15. M. Grobosch, C. Schmidt, R. Kraus, M. Knupfer, Electronic properties of transition metal phthalocyanines: the impact of the central metal atom (d5–d10). Org. Electron. **11**(9), 1483–1488 (2010). doi:10.1016/j.orgel.2010.06.006
16. T.G. Gopakumar, T. Brumme, J. Kröger, C. Toher, G. Cuniberti, R. Berndt, Coverage-driven electronic decoupling of Fe-phthalocyanine from a Ag(111) substrate. J. Phys. Chem. C **115**(24), 12173–12179 (2011). doi:10.1021/jp2038619
17. N. Tsukahara, S. Shiraki, S. Itou, N. Ohta, N. Takagi, M. Kawai, Evolution of Kondo resonance from a single impurity molecule to the two-dimensional lattice. Phys. Rev. Lett. **106**(18), 187201 (2011). doi:10.1103/PhysRevLett.106.187201
18. A. Zhao, Controlling the Kondo effect of an adsorbed magnetic ion through its chemical bonding. Science **309**(5740), 1542–1544 (2005). doi:10.1126/science.1113449
19. I. Fernández-Torrente, K.J. Franke, J.I. Pascual, Vibrational Kondo effect in pure organic charge-transfer assemblies. Phys. Rev. Lett. **101**(21), 217203 (2008). doi:10.1103/PhysRevLett.101.217203
20. N. Roch, C.B. Winkelmann, S. Florens, V. Bouchiat, W. Wernsdorfer, F. Balestro, Kondo effects in a C_{60} single-molecule transistor. Phys. Stat. Sol. (b) **245**(10), 1994–1997 (2008)
21. N. Roch, S. Florens, T.A. Costi, W. Wernsdorfer, F. Balestro, Observation of the underscreened Kondo effect in a molecular transistor. Phys. Rev. Lett. **103**(19), 197202 (2009). doi:10.1103/PhysRevLett.103.197202
22. T. Choi, S. Bedwani, A. Rochefort, C. Chen, A.J. Epstein, J.A. Gupta, A single molecule Kondo switch: multistability of tetracyanoethylene on Cu(111). Nano Lett. **10**(10), 4175–4180 (2010)
23. J. Nygaard, D.H. Cobden, P.E. Lindelof, Kondo physics in carbon nanotubes. Nature **408**(November), 342–342 (2000)
24. J. Paaske, A. Rosch, P. Wölfle, N. Mason, C.M. Marcus, J. Nygård, Non-equilibrium singlet-triplet Kondo effect in carbon nanotubes. Nat. Phys. **2**(7), 460–464 (2006)
25. E.A. Osorio, K. Moth-Poulsen, H.S.J. van der Zant, J. Paaske, P. Hedegård, K. Flensberg, J. Bendix, T. Bjornholm, Electrical manipulation of spin states in a single electrostatically gated transition-metal complex. Nano Lett. **10**(1), 105–110 (2009)
26. U. Perera, H. Kulik, V. Iancu, L. Dias da Silva, S. Ulloa, N. Marzari, S. Hla, Spatially extended Kondo state in magnetic molecules induced by interfacial charge transfer. Phys. Rev. Lett. **105**(10), 106601 (2010). doi:10.1103/PhysRevLett.105.106601
27. M. Liao, S. Scheiner, Electronic structure and bonding in metal phthalocyanines, Metal = Fe, Co, Ni, Cu, Zn, Mg. J. Chem. Phys. **114**(22), 9780 (2001). doi:10.1063/1.1367374
28. J. Bartolomé, F. Bartolomé, L.M. García, G. Filoti, T. Gredig, C.N. Colesniuc, I.K. Schuller, J.C. Cezar, Highly unquenched orbital moment in textured Fe-phthalocyanine thin films. Phys. Rev. B **81**(19), 195405 (2010). doi:10.1103/PhysRevB.81.195405
29. F. Roth, A. König, R. Kraus, M. Grobosch, T. Kroll, M. Knupfer, Probing the molecular orbitals of FePc near the chemical potential using electron energy-loss spectroscopy. Eur. Phys. J. B **74**(3), 339–344 (2010). doi:10.1140/epjb/e2010-00104-8
30. N. Marom, L. Kronik, Density functional theory of transition metal phthalocyanines, i: Electronic structure of NiPc and CoPc—self-interaction effects. Appl. Phys. A **95**(1), 159–163 (2008). doi:10.1007/s00339-008-5007-z
31. J. Sau, J.B. Neaton, H.J. Choi, S.G. Louie, M.L. Cohen, Electronic energy levels of weakly coupled nanostructures: C_{60}-metal interfaces. Phys. Rev. Lett. **101**, 026804 (2008)
32. A. Greuling, M. Rohlfing, R. Temirov, F.S. Tautz, F.B. Anders, Ab initio study of a mechanically gated molecule: From weak to strong correlation. Phys. Rev. B **84**(12), 125413 (2011). doi:10.1103/PhysRevB.84.125413

References

33. J. Göres, D. Goldhaber-Gordon, S. Heemeyer, M.A. Kastner, H. Shtrikman, D. Mahalu, U. Meirav, Fano resonances in electronic transport through a single-electron transistor. Phys. Rev. B **62**(3), 2188–2194 (2000). doi:10.1103/PhysRevB.62.2188
34. V. Mantsevich, N. Maslova, Different behaviour of local tunneling conductivity for deep and shallow impurities due to coulomb interaction. Solid State Commun. **150**(41–42), 2072–2075 (2010). doi:10.1016/j.ssc.2010.07.051
35. D.M. Kolb, W. Boeck, K. Ho, S.H. Liu, Observation of surface states on Ag(100) by infrared and visible electroreflectance spectroscopy. Phys. Rev. Lett. **47**(26), 1921–1924 (1981). doi:10.1103/PhysRevLett.47.1921
36. A.J. Cohen, P. Mori-Sánchez, W. Yang, Insights into current limitations of density functional theory. Science **321**(5890), 792–794 (2008). doi:10.1126/science.1158722
37. K. Nagaoka, T. Jamneala, M. Grobis, M.F. Crommie, Temperature dependence of a single Kondo impurity. Phys. Rev. Lett. **88**(7), 077205 (2002). doi:10.1103/PhysRevLett.88.077205
38. J.J. Parks, A.R. Champagne, T.A. Costi, W.W. Shum, A.N. Pasupathy, E. Neuscamman, S. Flores-Torres, P.S. Cornaglia, A.A. Aligia, C.A. Balseiro, G.K. Chan, H.D. Abruña, D.C. Ralph, Mechanical control of spin states in spin-1 molecules and the underscreened Kondo effect. Science **328**(5984), 1370–1373 (2010). doi:10.1126/science.1186874
39. B.C. Stipe, Single-molecule vibrational spectroscopy and microscopy. Science **280**, 1732–1735 (1998). doi:10.1126/science.280.5370.1732
40. N. Lorente, J. Gauyacq, Efficient spin transitions in inelastic electron tunneling spectroscopy. Phys. Rev. Lett. **103**(17), 176601 (2009)
41. M.N. Kiselev, Dynamical symmetries and quantum transport through nanostructures. Phys. stat. sol. (c) **4**(9), 3362–3373 (2007). doi:10.1002/pssc.200775419
42. D. Li, Z. Peng, L. Deng, Y. Shen, Y. Zhou, Theoretical studies on molecular structure and vibrational spectra of copper phthalocyanine. Vib. Spectrosc. **39**(2), 191–199 (2005)
43. N. Lorente, M. Persson, L.J. Lauhon, W. Ho, Symmetry selection rules for vibrationally inelastic tunneling. Phys. Rev. Lett. **86**(12), 2593–2596 (2001). doi:10.1103/PhysRevLett.86.2593
44. R. Korytár, N. Lorente, Multi-orbital non-crossing approximation from maximally localized wannier functions: the Kondo signature of copper phthalocyanine on Ag(100). J. Phys.: Condens. Matter **23**(35), 355009 (2011). doi:10.1088/0953-8984/23/35/355009
45. S. Sasaki, S. De Franceschi, J.M. Elzerman, W.G. van der Wiel, M. Eto, S. Tarucha, L.P. Kouwenhoven, Kondo effect in an integer-spin quantum dot. Nature **405**(6788), 764–767 (2000). doi:10.1038/35015509
46. M. Pletyukhov, D. Schuricht, H. Schoeller, Relaxation versus decoherence: spin and current dynamics in the anisotropic Kondo model at finite bias and magnetic field. Phys. Rev. Lett. **104**(10), 106801 (2010)
47. A. Rosch, J. Kroha, P. Wölfle, Kondo effect in quantum dots at high voltage: universality and scaling. Phys. Rev. Lett. **87**(15), 156802 (2001)
48. N. Roch, S. Florens, V. Bouchiat, W. Wernsdorfer, F. Balestro, Quantum phase transition in a single-molecule quantum dot. Nature **453**, 633–637 (2008)
49. J. Park, A.N. Pasupathy, J.I. Goldsmith, C. Chang, Y. Yaish, J.R. Petta, M. Rinkoski, J.P. Sethna, P.L. McEuen, D.C. Ralph, Coulomb blockade and the Kondo effect in single-atom transistors. Nature **417**(June), 722 (2002)
50. V.P. Zhukov, E.V. Chulkov, P.M. Echenique, First-principle approach to the study of spin relaxation times of excited electrons in metals. Phys. Stat. Sol. (a) **205**(6), 1296–1301 (2008)
51. A. Kogan, S. Amasha, M.A. Kastner, Photon-induced Kondo satellites in a single-electron transistor. Science **304**(5675), 1293–1295 (2004)
52. C. Buizert, A. Oiwa, K. Shibata, K. Hirakawa, S. Tarucha, Kondo universal scaling for a quantum dot coupled to superconducting leads. Phys. Rev. Lett. **99**(13), 136806 (2007)
53. S. Stepanow, A. Mugarza, G. Ceballos, P. Moras, J. Cezar, C. Carbone, P. Gambardella, Giant spin and orbital moment anisotropies of a Cu-phthalocyanine monolayer. Phys. Rev. B **82**(1), 014405 (2010). doi:10.1103/PhysRevB.82.014405
54. I. Kröger, B. Stadtmüller, C. Stadler, J. Ziroff, M. Kochler, A. Stahl, F. Pollinger, T. Lee, J. Zegenhagen, F. Reinert, C. Kumpf, Submonolayer growth of copper-phthalocyanine on Ag(111). New J. Phys. **12**, 083038 (2010). doi:10.1088/1367-2630/12/8/083038

55. P. Wahl, L. Diekhöner, G. Wittich, L. Vitali, M.A. Schneider, K. Kern, Kondo effect of molecular complexes at surfaces: ligand control of the local spin coupling. Phys. Rev. Lett. **95**(16), 166601 (2005). doi:10.1103/PhysRevLett.95.166601
56. V. Iancu, A. Deshpande, S.-w. Hla, Manipulating Kondo temperature via single molecule switching. Nano Lett. **6**, 820–823 (2006). doi:10.1021/nl0601886
57. J. Kröger, L. Limot, H. Jensen, R. Berndt, P. Johansson, Stark effect in Au(111) and Cu(111) surface states. Phys. Rev. B **70**(3), 033401 (2004). doi:10.1103/PhysRevB.70.033401
58. Y.C. Choi, W.Y. Kim, K. Park, P. Tarakeshwar, K.S. Kim, T. Kim, J.Y. Lee, Role of molecular orbitals of the benzene in electronic nanodevices. J. Chem. Phys. **122**(9), 094706 (2005). doi:10.1063/1.1858851
59. R. Temirov, S. Soubatch, A. Luican, F.S. Tautz, Free-electron-like dispersion in an organic monolayer film on a metal substrate. Nature **444**(7117), 350–353 (2006). doi:10.1038/nature05270
60. M.S. Dyer, M. Persson, The nature of the observed free-electron-like state in a PTCDA monolayer on Ag(111). New J. Phys. **12**(6), 063014 (2010). doi:10.1088/1367-2630/12/6/063014
61. J. Zhao, C. Zeng, X. Cheng, K. Wang, G. Wang, J. Yang, J.G. Hou, Q. Zhu, Single $C_{59}N$ molecule as a molecular rectifier. Phys. Rev. Lett. **95**(4), 045502 (2005b). doi:10.1103/PhysRevLett.95.045502
62. B. Li, C. Zeng, J. Zhao, J. Yang, J.G. Hou, Q. Zhu, Single-electron tunneling spectroscopy of single C_{60} in double-barrier tunnel junction. J. Chem. Phys. **124**(6), 064709 (2006). doi:10.1063/1.2163333
63. G.V. Nazin, S.W. Wu, W. Ho, Tunneling rates in electron transport through double-barrier molecular junctions in a scanning tunneling microscope. Proc. Natl. Acad. Sci. U S A **102**(25), 8832–8837 (2005). doi:10.1073/pnas.0501171102
64. S.W. Wu, G.V. Nazin, X. Chen, X.H. Qiu, W. Ho, Control of relative tunneling rates in single molecule bipolar electron transport. Phys. Rev. Lett. **93**(23), 236802 (2004). doi:10.1103/PhysRevLett.93.236802
65. M. Grobis, A. Wachowiak, R. Yamachika, M.F. Crommie, Tuning negative differential resistance in a molecular film. Appl. Phys. Lett. **86**(20), 204102 (2005). doi:10.1063/1.1931822
66. X.H. Qiu, G.V. Nazin, W. Ho, Vibronic states in single molecule electron transport. Phys. Rev. Lett. **92**(20), 206102 (2004). doi:10.1103/PhysRevLett.92.206102
67. T. Schwieger, H. Peisert, M. Golden, M. Knupfer, J. Fink, Electronic structure of the organic semiconductor copper phthalocyanine and K-CuPc studied using photoemission spectroscopy. Phys. Rev. B **66**(15), 155207 (2002). doi:10.1103/PhysRevB.66.155207
68. I. Hill, Charge-separation energy in films of ϕ-conjugated organic molecules. Chem. Phys. Lett. **327**(3–4), 181–188 (2000). doi:10.1016/S0009-2614(00)00882-4
69. X.W. Tu, G. Mikaelian, W. Ho, Controlling single-molecule negative differential resistance in a double-barrier tunnel junction. Phys. Rev. Lett. **100**(12), 126807 (2008). doi:10.1103/PhysRevLett.100.126807
70. I. Fernández Torrente, K.J. Franke, J. Ignacio Pascual, Spectroscopy of C_{60} single molecules: the role of screening on energy level alignment. J. Phys.: Condens. Matter **20**(18), 184001 (2008). doi:10.1088/0953-8984/20/18/184001
71. R. Hesper, L.H. Tjeng, G.A. Sawatzky, Strongly reduced band gap in a correlated insulator in close proximity to a metal. Europhys. Lett. **40**(2), 177–182 (1997). doi:10.1209/epl/i1997-00442-2
72. J. Repp, G. Meyer, Scanning tunneling microscopy of adsorbates on insulating films. From the imaging of individual molecular orbitals to the manipulation of the charge state. Appl. Phys. A **85**(4), 399–406 (2006). doi:10.1007/s00339-006-3703-0
73. S. Kera, H. Yamane, N. Ueno, First-principles measurements of charge mobility in organic semiconductors: valence hole-vibration coupling in organic ultrathin films. Prog. Surf. Sci. **84**(5–6), 135–154 (2009). doi:10.1016/j.progsurf.2009.03.002

74. S. Kera, H. Yamane, I. Sakuragi, K.K. Okudaira, N. Ueno, Very narrow photoemission bandwidth of the highest occupied state in a copper-phthalocyanine monolayer. Chem. Phys. Lett. **364**(1–2), 93–98 (2002). doi:10.1016/S0009-2614(02)01302-7
75. F. Evangelista, V. Carravetta, G. Stefani, B. Jansik, M. Alagia, S. Stranges, A. Ruocco, Electronic structure of copper phthalocyanine: an experimental and theoretical study of occupied and unoccupied levels. J. Chem. Phys. **126**(12), 124709 (2007). doi:10.1063/1.2712435

Chapter 6
Doping of MePc: Alkali and Fe Atoms

In this chapter we will explore the possibility to manipulate the charge and spin states of MePcs by adding dopant adatoms. To electron dope the molecules, we codeposited them with a known electron donor, lithium (Li) atoms. In a second experiment, we studied the interaction of NiPc and CuPc, with magnetic Fe adatoms as a more direct way to influence the molecular magnetic moment.

6.1 Electron Doping of MePc

In "traditional" electronics the doping of semiconductors is a common approach to achieve a desired electronic structure. This idea has been carried over to molecular electronics. Consequently considerable research has been done on electron/hole doping of organic films.

In purely organic films of simple molecules such as C_{60} the scenario is rather simple. Electron doping generally leads to a rigid, continuous, multiple filling of the unoccupied π orbitals [1–3]. On the other hand MePcs, with localized d states coexisting with π orbitals near E_F, reveal a more complex behavior. Previous experiments on MePc thin films (50–100 nm) conclude that the doping depends mainly on the type of metal ion. Studies on potassium (K) doped thin films of CoPc and FePc show that the charging occurs via the ion's d orbitals [4–6]. In contrast, for NiPc and CuPc molecules, the charging of the ligand LUMO is reported [4, 5, 7]. DFT calculations suggest that this behavior is due to the presence or absence of unfilled/partially filled d states close to E_F, as well as the metal ion's on-site coulomb repulsion, as we have seen in the previous chapter [8]. Furthermore, in these thin films the crystal structure is deeply intertwined with the charging behavior. In K doped MePc crystals (Me = Fe [9], Cu [10], Zn [11]) a phase separation between 2K@MePc and 4K@MePc occurs. Each different structure allows a transfer of respectively 2 or 4 electrons to the molecule. A recent study of a CuPc monolayer on Au(110) shows the charging can go even further and involve not only the LUMO, but the LUMO+1 and LUMO+2, thus accepting more charge per CuPc [12]. The type of dopant used has an effect as well. For instance sodium and rubidium doped CuPc films report the charging of the Cu ion without any phase separation [13, 14].

Nevertheless, the doping of single MePcs at a metallic interface has not been studied so far, although it lies very well within the capabilities of an STM setup, as shown by the controlled doping of C_{60} with potassium [3]. The results presented in the previous chapters show that, already for undoped molecules, a range of intriguing electronic and magnetic states can be achieved by charge transfer to MePc with different metal ions. The atom-by-atom doping with Li atoms is the next step to further extend and tune the MePc's rich magnetic behavior.

In this section, the results of a comparative study of the Lithium doping of individual MePcs (Me = Fe, Co, Ni, Cu). We will show that the charging behavior of these molecules adsorbed on an Ag(100) surface depends not only on the type of the central metal ion, but also on the specific bonding configuration of the Li and molecule, allowing the specific charging of ion or ligand states. In this fashion the total spin of individual CuPc molecules can be controlled from 0 to $1/2$ and 1, and the molecular charge from 1 to 2.

6.1.1 Single MePc Doped with Lithium

Topography of Li@CuPc and Li@NiPc

Li atoms were deposited on at 5 K on a surface containing single MePc molecules. The interaction between CuPc with Li atoms leads to several stable configurations. The two most frequent (more than 75% of all Li@CuPc) are presented in Fig. 6.1, denoted as L, M.

The type L (ligand) configuration exhibits two very bright protrusions on one of the benzene rings at negative bias voltage. From STM manipulation of Li atoms (see the discussion of Fig. 6.9 on page 125) it becomes clear that this is the effect of one Li atom interacting with one of the benzene rings of the CuPc. When two Li atoms interact with a CuPc, the bright protrusions are seen on two opposing benzene rings, as in the 2L type. Other configurations, such as Li adatoms in two neighboring benzene rings were not observed.

The type M (metal) configuration leads to a Li@CuPc with fourfold symmetric ligands, and a brighter center than the undoped CuPc, suggesting that one Li atom is located at the center of the molecule. The rotational stability is affected by the Li, and some of the type M molecules are rotated $+14°$; $-42°$ with respect to the [011] direction, whereas for the type L complexes no deviation from the $\pm 30°$ orientation of the undoped CuPc were observed.

For Li@NiPc only the type L configuration is stable. A configuration that could be assigned to type M appears to be unstable, with a fuzzy ion site both in topography and spectroscopy (data not shown). The lower stability of this configuration in NiPc is also supported by the DFT calculations, as will be treated later on.

6.1 Electron Doping of MePc

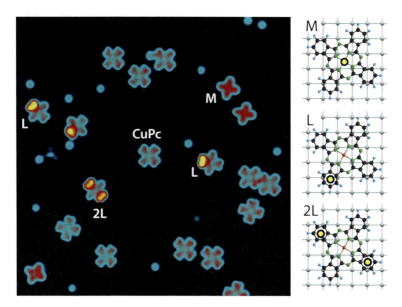

Fig. 6.1 STM topography (−0.3 V, 0.17 nA, 19.3 nm × 19.3 nm) of CuPc and Li atoms on Ag(100). Different types of xLi@CuPc complexes are marked with L and M. Their proposed geometrical configurations are shown on the *right panel*

Electronic Structure of Li@CuPc and Li@NiPc: Orbital Selectivity

To characterize the influence of the Li on the electronic structure, STS measurements were performed. Due to the low sensitivity of the tunneling current to the ion's *d*-states near the Fermi level, especially if the Li atom sits on top of the metal ion, we restrict our spectroscopic study to the dI/dV signal measured on the benzene rings, as shown in Fig. 6.2.

In gas phase MePc molecules the empty $2e_g$ state is 2-fold orbital degenerate. For the CuPc/Ag(100) it is singly occupied. We observe a peak below E_F, and an additional peak above E_F corresponding to the single ($2e_g$) and double occupation ($2e_g$+U) respectively (see Chap. 5). For both Li@CuPc types the $2e_g$ orbital is in a different configuration. In dI/dV spectra of type M complexes a single unoccupied state is encountered. Its spatial distribution follows that of the $2e_g$, indicating that the effect of the Li is to deplete it. In type L complexes a double peak structure is recovered. In this case the degeneracy of the $2e_g$ state is lifted, and the MO is split into two separate states, denoted as α_{2e_g} (occupied state) and β_{2e_g} (unoccupied state). This can be deduced from the twofold symmetric, orthogonal molecular resonances mapped at each energy, as opposed to the fourfold symmetric orbital mapped for both undoped and type M complexes, as shown in Fig. 6.2. The reason for this is the symmetry reduction from C_{4v} to C_{2v} caused by the presence of Li, similar to the effect induced by a neighboring molecule discussed on page 93. Note that the

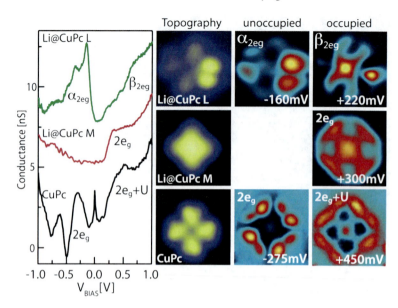

Fig. 6.2 dI/dV spectra taken on one benzene ring of different xLi@CuPc complexes (1.0 nA, −1.0 V) and CuPc on Ag(100) (3.0 nA, −2.0 V). *Right panels*: dI/dV maps 3.0 nm × 3.0 nm, showing the intensity distribution of the marked features in the spectra. Note that for the L type the α_{2eg} and β_{2eg} states are orthogonal to each other

intense double lobe structure of the α_{2eg} state at the benzene ring is responsible for the topographic appearance at negative bias voltage.

In both Li@CuPc species the Kondo resonance is quenched, due to the changes in the $2e_g$ state. For type M this is readily explained from the dI/dV spectra by the loss of the unpaired electron in the $2e_g$ orbital. The situation for type L is the opposite. The fact that we do not observe the double occupation Coulomb pair of α_{2eg} above E_F, but only the split β_{2eg} state, suggests that the α_{2eg} orbital is completely filled, with a charge state of $Q = 2$. This interpretation is confirmed by DFT as shown later, which reveals that the double occupation consists of one electron donated from the Ag substrate and another from the Li atom (see Table 6.1).

Despite the similar topographic appearance, the charging behavior of Li@NiPc differs from that of Li@CuPc. Figure 6.3 shows spectra for the undoped, singly and doubly doped NiPc molecules. The spectrum of 1Li@NiPc-L is different from that of its CuPc counterpart, presenting sharp peaks at both sides of E_F. The dI/dV maps, taken at the energy of these peaks, show both a node at the axis intersecting the benzene with Li, indicating that they correspond to the same orbital, the single and double occupation of the α_{2eg} orbital. The orthogonal β_{2eg} state is found at slightly higher energy, indicating a completely empty orbital. A weak broad hump overlapping with the α_{2eg}+U state in the spectra can be tentatively associated to this orbital. Based on the double peak structure around E_F observed for the α_{2eg} orbital,

6.1 Electron Doping of MePc

Table 6.1 LDA-DFT calculations, for 6 different positions of the Li atom. The formation energy ΔE is given with respect to the most stable configuration

		L_1	L_2	L_3	L_4	M_1	M_2
ΔE [eV]	CuPc	0.22	0.49	0.88	0.87	0	0.03
	NiPc	0	0.31	0.67	0.71	0.53	0.99
$m[\mu_B]$	CuPc	0.44	0.42	0.45	0.42	0	0
	NiPc	0	0	0	0	0	0.79
ΔN (MePc)	CuPc	1.58	1.44	1.85	1.98	1.98	2.06
	NiPc	1.38	1.27	1.69	1.8	1.9	2.02
ΔN (TM)	CuPc	0.08	0.08	0.09	0.11	0.47	0.66
	NiPc	0.04	0.04	0.05	0.07	0.15	0.3
ΔN (Li)	CuPc	−0.85	−0.81	−0.87	−0.87	−0.87	0.96
	NiPc	−0.89	−0.80	−0.86	−0.84	−0.88	−0.89
ΔN (Ag)	CuPc	−0.73	0.63	−0.98	−1.11	−1.12	−1.1
	NiPc	−0.49	−0.47	−0.83	−0.96	−1.02	−1.13

m is the magnetic moment of the ion (units of μ_B). ΔN denotes the charge difference

similar to that found for the $2e_g$ for the undoped molecule, we conclude that this state is singly occupied.

These observations imply that Li@NiPc-L has only one added electron (Q = 1), as opposed to the two found for Li@CuPc-L (Q = 2). The lower charge state for Li@NiPc is confirmed by adding a second Li atom. The spectra of 2Li@NiPc-L becomes identical to that of Li@CuPc-L. In this case the orthogonality of the states found above and below E_F is even more clear, confirming the presence of a filled α_{2eg} and an unoccupied β_{2eg} state (see Fig. 6.3a).

This difference in the charging behavior for NiPc and CuPc is also seen in gas-phase electron affinity calculations. [CuPc]$^{1-}$ with an electron affinity of EA = 0.27 eV [1] accepts an additional electron more easily than [NiPc]$^{1-}$ with EA = 0.8 eV [8]. Nevertheless energy difference between the Q = 1 and Q = 2 charge state for the adsorbed molecules is not very large. The α_{2eg} state in NiPc can be doubly occupied by using a more electropositive alkali such as K. This can be seen in Fig. 6.4b. The dI/dV spectrum taken on the K@NiPc-L complex is very similar to that of Li@CuPc-L and 2Li@NiPc-L. On the other hand by slightly changing the adsoprtion configuration of Li@CuPc-L, the charge state can be reduce from 2 to 1, as shown in Fig. 6.4b. Here the position of the Li adatom also seems to be at one of the benzene rings of the molecule, although the two bright lobes appear slightly

---
[1] Positive values mean an additional energy cost.

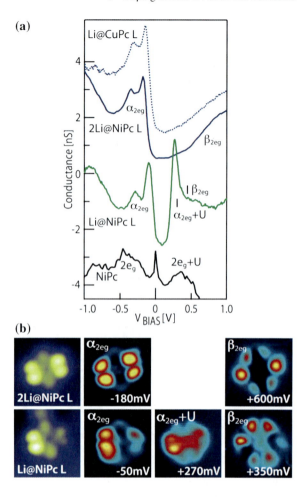

Fig. 6.3 a dI/dV spectra taken on one of the benzene rings of different xLi@NiPc and an undoped NiPc (1.0 nA, −1.0 V). The undoped NiPc/Ag(100) spectra is background subtracted for clarity. **b** STM topography and dI/dV maps 3 nm × 3 nm, showing the charged and uncharged MO

different as in the standard type L configuration. Furthermore, the azimuthal angle with respect to the [011] axis is 50° rather than the 30° found for the type M and L complexes. The spectroscopy taken on the benzene ring of this complex is similar to the Li@NiPc-L type, showing sharp features above and below E_F. We hence conclude that both of these complexes have a total molecular charge of $Q = 1$.

Vibronic Coupling

We now focus on the multi-peak structure found in the occupied states of both the L type Li@CuPc and Li@NiPc complexes. Based on the equidistant energy spacing of the individual peak features, we assign it to the coupling of the α_{2eg} orbital to vibrational modes of the molecule (see Fig. 6.5). This effect has also been observed in CuPc multilayers, as discussed in Sect. 5.5.1, (page 101). The measured energy spacing of 173 mV and 193 mV lies close to that found in multilayers, and within

6.1 Electron Doping of MePc

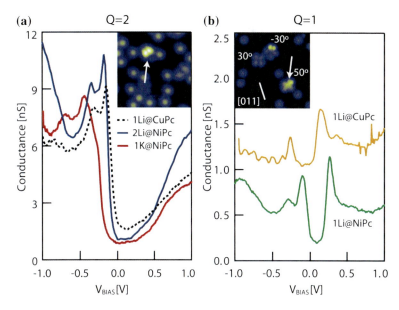

Fig. 6.4 **a** dI/dV spectra for 1Li@CuPc-L, 2Li@NiPc-L, and 1K@NiPc-L. The similar spectra indicate that by using a more electropositive dopant such as K, we can charge NiPc to $Q=2$. **b** 1Li@CuPc molecule in a rare configuration, with ligand axes rotated 50° with respect to the surface symmetry axes [011], 20° more than undoped CuPc and the L and M species. The spectra is very similar to 1Li@NiPc-L, indicating a molecular charge of $Q=1$

the range reported for C–C or C–N stretching modes in literature (150–200 meV) [15–17]. Their existence is signature of a decoupling from the conduction electrons of the substrate.

DFT: MePc with Li Atoms

STS was able to detect the charging trend of the ligand p orbital for the different Li@MePc configurations, however the behavior of TM d states cannot be accessed as readily. Hence, the full electronic and magnetic structure of the charged molecules can only be understood with the aid of DFT calculations.

Six different configurations, depicted in Fig. 6.6c, have been explored by DFT. In four of them the Li ion interacts with the organic ligand, and in the other two with the metal ion. The relative energies plotted in Fig. 6.6b, show a clear switch in the stability of configurations with Li interacting with the different ions (M_1, M_2): for CuPc they are stable configurations, whereas for NiPc they are less stable. This explains the fuzzy images obtained for Li@NiPc with the Li at the center.

The configurations that involve a Li-ligand interaction show similar trends for both molecules: the most stable one places the Li outside the molecule close to the aza-N (L_1), and the second most stable has the Li under a benzene ring (L_2). Next are the positions over the benzene ring (L_3), and over the pyrrole group (L_4). All four

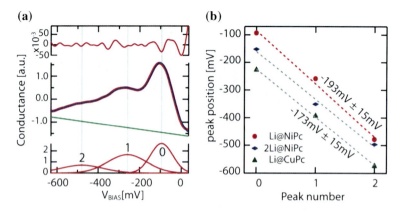

Fig. 6.5 **a** Fit of the multi peak structure observed below E_F for Li@NiPc. The *above curve* shows the error of the fit, and the fitted peaks are presented. **b** Peak position obtained from similar fits for different Li@MePc configurations. An equidistant peak spacing is visible, pointing to a vibronic origin of the multi-peak structure

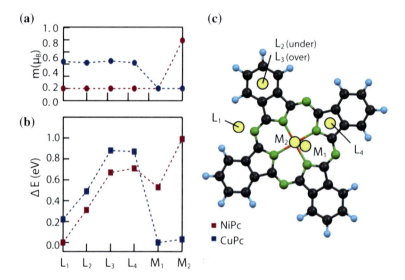

Fig. 6.6 DFT calculations for various positions of the Li adatom: **a** Magnetic moment m (units of μ_B) of the TM ion, and **b** relative formation energies ΔE for various positions of the Li, presented in (**c**). Note that for the M_1 configuration the charge is transferred to the $2e_g$ for NiPc whereas for CuPc to the transfer is to the b_{1g} orbital (see Table 6.1)

configurations, considering the Li-ligand bond, show the same electronic structure, regardless of the exact position of the Li.

We tentatively assign the type L complex to the configuration with the Li adatom underneath the benzene ring (L_2), because: (i) the STM images show a clear intense signal coming from one of the benzene rings in these molecules, (ii) no direct topographic indication of a Li adatom on top of the ring is observed, and (iii) the DFT

6.1 Electron Doping of MePc 123

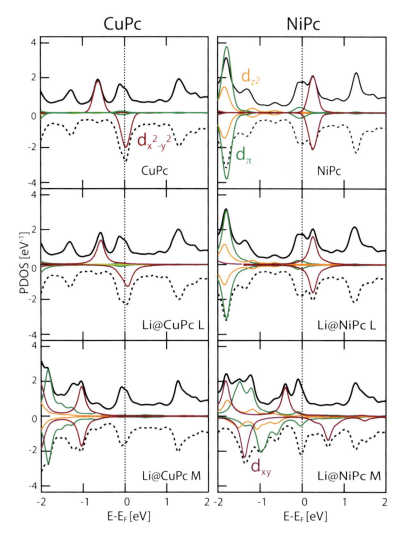

Fig. 6.7 LDA-DFT calculations: spin resolved DOS projected on the d states of the metal ion for NiPc and CuPc molecules with the Li atoms under the benzene ring (L$_2$) and on *top* of the central ion (M$_2$)

calculations find it to be more stable than the configuration with the Li positioned over the benzene ($\Delta E = 360$ meV for NiPc and $\Delta E = 390$ meV for CuPc).

The calculated PDOS for the Li@MePc complexes is shown in Fig. 6.7. Both for CuPc and NiPc the configurations with Li-ligand bonding, represented by L$_3$, show a downshift of the $2e_g$ state further below E_F as compared to the undoped molecules, indicating the additional charging of this state. The molecule accepts between 1 and 2 electrons, with the exact number slightly varying for each configuration (see Table 6.1). The transferred charge originates from both the substrate and the alkali

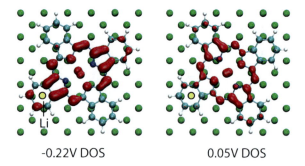

Fig. 6.8 Calculated charge density isosurfaces (LDA-DFT) of the Li@NiPc/Ag(100) system for the occupied α_{2eg} and unoccupied β_{2eg} states. The position of the Li atom is over the benzene ring

atom. The electronic structure of the metal ion remains unchanged, with a single electron in the b_{1g} ($d_{x^2-y^2}$) orbital in CuPc and none in NiPc. Although the lifting of the $2e_g$ degeneracy should be present in all cases of Li-ligand bonding it is best seen for Li on top of the benzene ring (L$_3$), probably due to an overlapping of the weakly split peaks in the other cases (see Fig. 6.8). In general these results are in good agreement with experimental observations.

The two configurations related to the metal ion reveal a continuous transition of charge transfer to the metal ion b_{1g} state. When the Li adatom is slightly shifted from the TM position towards the aza-N, the transferred charge is still located in the $2e_g$ state in NiPc, but is already accepted by the b_{1g} state in CuPc. This again demonstrates the higher affinity of CuPc to bond via the metal ion d states. In the M$_2$ configuration the charge is transferred to the b_{1g} state in both molecules. Based on the symmetry of the type M complex, we assign the latter to this configuration.

The calculations do not fully reproduce the offset in the charging behavior of CuPc and NiPc, however the general tendency of the latter to accept less charge can be seen in Table 6.1. The charge transfer from the Ag substrate seems to be suppressed by the presence of Li. For undoped MePcs roughly 1 electrons was transferred from the Ag (Table 5.1 on page 90), whereas for Li@MePc-L complexes it is less than 1. The total transfer is generally less than 2 electrons in both molecules. This value is always smaller for Li@NiPc, closer to 1 than to 2 in the L$_1$ and L$_2$ configuration, in line with the experimentally observed delayed charging compared to CuPc.

The difference between the experimental and theoretical charge transfer could be explained by the underestimated interorbital Coulomb repulsion in LDA-DFT calculations. The experimentally observed single charging in Li@NiPc-L, and the the depletion of the $2e_g$ in M type complexes might be stabilized by the higher Coulomb repulsion present in the experiment.

6.1.1.1 Atom-by-Atom Doping of CuPc

By using the manipulation capabilities of the STM to move individual atoms, we have created xLi@CuPc complexes with up to 5 Li atoms. Figure 6.9b shows the topographies, obtained after the addition of Li to the complex. The first two Li atoms are localized on opposite benzene rings in the type L configuration. With the

6.1 Electron Doping of MePc

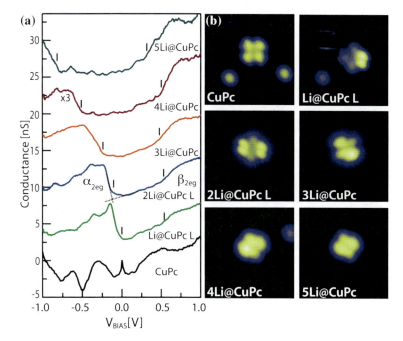

Fig. 6.9 a dI/dV spectra taken on the benzene rings on different xLi@CuPc on Ag(100) (1.0 nA, −1.0 V). **b** STM topography (0.23 nA, −0.3 V, 6 nm × 6 nm) showing the manipulation steps to add 1–5 Li atoms to a single CuPc

introduction of the third Li, however, it becomes difficult to determine the exact position of the alkali ions. In any case the topography suggest that all of them are located in the two bright opposing benzene rings.

Charge Saturation

In the dI/dV spectra taken on the benzene rings, we observe the symmetry induced splitting of the $2e_g$ and the filling of the α_{2eg} orbital upon the doping with the first Li atom. After the addition of more Li, no further Fermi level crossing of any peaks, but rather a continuous downshift of the α_{2eg} orbital is observed. This behavior indicates that the charge transfer to the ligand orbitals is saturated. We attribute this to the splitting of the $2e_g$ orbital, which seems to open a high enough α_{2eg}–β_{2eg} gap, making it energetically unfavorable to add more electrons to the ligand. Note that from this experiment we cannot exclude the transfer of one additional electron to the single d-hole in the Cu ion, although in energy calculations of gas-phase MePc anions this orbital starts to fill only after the completion of the $2e_g$ [8].

Our result stands in clear contrast to the observed charging up to the LUMO+2 of a CuPc monolayer on Au(110) doped with K [12]. This could be due to the decreased ionization energy of K compared to Li, which stabilizes the larger charge transfer, as we have seen for NiPc. However the geometrical agreement of the dopant atoms could also be the cause, as seen in K doped thin MePc films (100 nm). Here a phase separation for 2K@CuPc and 4K@CuPc phases, with two different geometrical arrangements, results in Q = 2 and Q = 4 configurations [10]. The higher stabilization of Q = 2 in the isolated molecule studied could be related to the fact that all Li atoms bonded to the molecule are positioned in one ligand axis, minimizing the interaction with β_{2eg}.

The short-range electrostatic interaction of the Li ions within the molecule seems to stabilize the Q = 2 state by increasing the energy gap via electrostatic interactions. The α_{2eg} orbital exhibits an almost linear decrease in energy with the number of alkali ions as expected from electrostatic interactions with a positive charge, whereas the β_{2eg} orbital remains mostly unaffected. As a result the gap increases by a factor of 2, from 0.55 to 1.17 eV as measured using the onset of the peaks (see Fig. 6.10a). Such strong variations could play an important role in the homogeneity of the energy gap of doped organic semiconductor films.

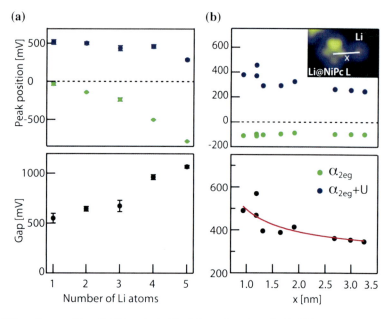

Fig. 6.10 a Onsets of the dI/dV peaks in Fig. 6.9 as a function of added Li atoms for xLi@CuPc (x = 1–5). The resulting gap opens linearly with additional Li atoms. **b** Energy position of the α_{2eg}, and α_{2eg}+U orbitals for Li@NiPc in relation to the distance of the nearest Li atom to the center of the molecule. The corresponding gap width can be fitted with a 1/x relation. Inset: STM topography of a Li@NiPc with a Li adatom near

6.1 Electron Doping of MePc

Long-Range Electrostatic Interaction

For the Li@NiPc-L we find a different electrostatic effect. Here the intraorbital Coulomb repulsion of α_{2eg} orbital seems to be sensitive to the nearby non-interacting Li ions. The energy gap between the peaks corresponding to the singly and doubly occupied α_{2eg} orbital increases as for decreasing distances x between the center of the complex and the nearest Li adatom (see Fig. 6.10b). This increase of the repulsion term for shorter molecule-ion distances can only be associated with an effective negative charge in the proximity, thus the effect seems to be more related to the charge redistribution induced around the Li ion rather than to its positive charge. This picture is in line with the 1/x behavior and the decay length of about 3 nm [2] observed in the energy gap variation, which is in the order of the electrostatic interactions observed for Li adatoms on Ag(100) [18] and the lateral extension of charge densities induced by positive charges on metallic surfaces [19].

6.1.1.2 Li Doping of FePc and CoPc

To further understand the effect of the central ion, the cases of the paramagnetic molecules FePc and CoPc were also investigated. We expect a drastically different behavior due to the ion's out-of-plane d states lying very close to E_F.

The addition of Li gives rise to only one stable configuration each for FePc and CoPc. The topography of both complexes are very similar to the undoped molecules. They can however be distinguished by a height difference in STM topography (Fig. 6.11) and an asymmetry of the central protrusion. We assign the position of the Li atom close to the center of the molecule, similar to the type M configuration, although the asymmetric shape of most complexes suggests that slight lateral shifts from the location of the central ion are possible (see Fig. 6.12a).

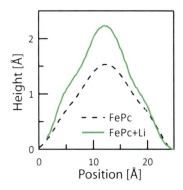

Fig. 6.11 STM topographic height profile for the Li@FePc complex and for undoped FePc/Ag(100) (0.1 nA, −1 V)

---

[2] This distance could be reduced by taking the distance to the periphery of the molecule instead of the center by as much as 0.5 nm.

128 6 Doping of MePc: Alkali and Fe Atoms

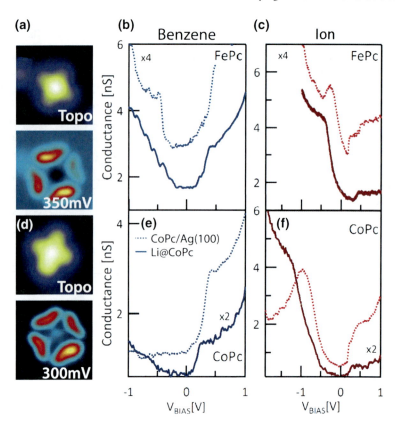

Fig. 6.12 **a** STM topography (1.0 nA, −0.5 V) and dI/dV map (1.1 nA, +0.35 V) of Li@FePc showing the uncharged $2e_g$. dI/dV spectra taken on the ion (**b**) benzene (**c**) of Li@FePc (*solid lines*) spectra taken on the undoped FePc are shown as reference (*dotted lines*) (1.0 nA, −1.0 V). **d** STM topography (0.36 nA, −1 V) and dI/dV map (0.5 nA, +0.3 V) for Li@CoPc. dI/dV spectra taken on the ion (**e**) and benzene (**f**) of Li@CoPc (0.2 nA, −1.0 V) with spectra for the undoped CoPc as a reference (1.0 nA, −1.0 V)

Both complexes exhibit a deviation from the regular azimuthal adsorption angle: We observe ±19° and −37° orientation for Li@FePc, and ±30°, +40°, +14°, −2°, −12°, −22° for Fe@CoPc. The changes in the orientation itself could contribute in the modifications observed in the electronic structure. In any case we observe only a few differences between the undoped and the doped molecules in STS measurements and no impact of the azimuthal orientation.

The spectroscopy on the benzene rings (Fig. 6.12c, f) shows a small downshift of the $2e_g$, which lies still above E_F for both FePc and CoPc. The dI/dV maps taken at the position of these peaks clearly follow the symmetry of the $2e_g$, confirming the origin of these peaks (compare to Fig. 5.3 on page 79). On the TM, the peak corresponding to the $d_\perp$ states experiences a comparable downshift in both cases (Fig. 6.12c,f), similar to that found for the $2e_g$ orbital. For FePc it still crosses the

Fermi level, which implies a situation with an uncompleted filling of 3d orbitals. For CoPc the peak is already completely below E_F and shifts from -1 to -1.3 V.

The DFT results for the undoped FePc/CoPc in the previous chapter reveal a complex mixed-valence situation. The molecular orbitals are hybridized with substrate states, and at the same time the ion based e_g and ligand based $2e_g$ are strongly intermixed. For single Li@FePc and Li@CoPc on Ag(100) it is thus difficult to assign charge transfer to specific orbitals.

6.1.2 Li Doping of a Monolayer of CuPc

We also studied the effect of Li doping on a complete monolayer of CuPc molecules adsorbed on Ag(100). The CuPc arrange in a homochiral close-packed array, with all molecules on each terrace oriented in the same way (see the description in Sect. 4.3.2 on page 61). The addition of a small amount of Li atoms leads to the creation of two different geometrical arrangements, analogue to the M and L structures found for single CuPc (Fig. 6.13a, b). We will keep the same nomenclature. The type M has a bright center insinuating that the Li atom is positioned over the Cu^{2+} ion. The topography for the type L on the other hand is characterized by two brighter isoindole ligands, suggesting that the Li atom might be bonded to the aza-N.[3]

In the dI/dV curves we find a different charging behavior depending on the position of the Li atom, comparable to the single molecules complexes. In Fig. 6.13c, d the spectra taken on the benzene ring are shown in comparison to those of undoped molecules within the layer. In general the charging behavior matches that of the single Li@CuPc.

The type M spectra show an unoccupied $2e_g$ state above E_F. Following the analogy to the single Li@CuPc-M molecule, we propose that the Li transfers one electron to the d states of the Cu ion, reducing its magnetic moment to zero. This picture is supported by our XMCD results presented later. For type L complexes the presence of Li induces a splitting of the $2e_g$ orbital again. This can be seen in Fig. 6.13d: the charged α_{2eg} state has its intensity only on one of bright the isoindole groups, whereas the unoccupied β_{2eg} is localized on the neighboring darker isoindole, revealing an orthogonal relationship. We can thus conclude that the $2e_g$ state is split and has accepted one electron from the Li atom. Further from the absence of the unoccupied Coulomb pair and the complete Fermi level crossing we can deduce that the α_{2eg} is doubly occupied.

These two configurations are however not equally stable. It is possible, by applying a voltage pulse (~ -2.2 V), to irreversibly switch the type M to a type L, as can be seen in Fig. 6.13a, b. This indicates that the type M complex by itself is less stable than type L. On the other hand, as we increase the density of Li atoms this stability criteria is reversed. Figure 6.14c shows the number of type L and M complexes found, as a function of Li coverage. At low Li dosage both types of configurations can be found

---

[3] Note that the aza-N bridges two isoindole groups.

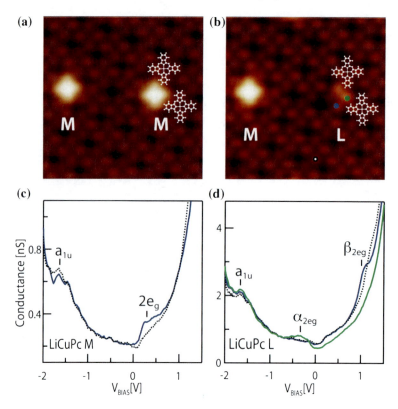

Fig. 6.13 a STM topography of two Li@CuPc-M complexes within a CuPc monolayer (0.17 nA, +0.2 V, 8.3 nm × 8.3 nm). **b** Topography of the same region after applying a voltage pulse (~ −2.2 V) on the right M type Li@CuPc molecule converting it into the L type. The positions of the dI/dV spectra shown in (**c**) are indicated. dI/dV spectra taken on the benzene rings of the Li@CuPc-M complex (**c**), and the Li@CuPc-L complex (**d**). The spectra of an undoped CuPc within the layer is shown as a *dotted line* (0.5 nA, −2.0 V)

on the surface. The more stable L type is slightly more frequent. With increasing coverage, however, we observe an overall increase of the type M configuration, which eventually becomes dominant over the L type. A possible driving force for this transition could be the electrostatic repulsion between individual Li ions.

This stabilization of charge transfer to the metal ion is in agreement with our XMCD data measured on heavily doped monolayers of CuPc and NiPc, where we can specifically measure the electronic valence and magnetic moment of the metal ion (see Fig. 6.15). In both cases we see the transfer of an additional electron to the ion's d states. The undoped CuPc layer shows a clear magnetic response from the Cu ion, which vanishes upon Li doping. For NiPc layers, we observe the opposite effect, the addition of Li leads to magnetic Ni ion. Note that this implies that the Li@NiPc type M configuration, which was unstable for single molecules, seems to

6.1 Electron Doping of MePc

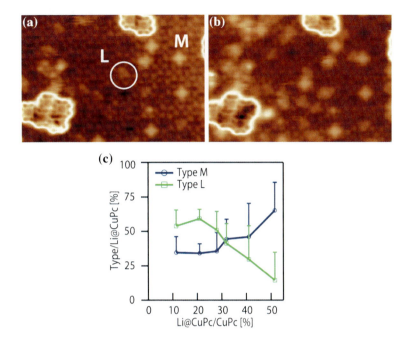

Fig. 6.14 a STM topography for a CuPc monolayer with a low Li coverages. Residual molecules in the 2nd layer are shown in reverse contrast (0.17 nA, +0.2 V, 18.5 nm × 12.7 nm). **b** Topography of the same region with higher Li coverage. **c** Statistics of the type of complex as a function of Li coverage, extracted from topographic images. Roughly 500 molecules were counted. Only molecules that were unequivocally identified were used. The error bars acount for undefined configurations

be stabilized by the high Li coverage, possibly by the same mechanism causing the crossover of stability found in doped CuPc monolayers.

6.2 Doping with Fe Atoms

We have seen that the magnetic structure of MePc is strongly influenced by the interaction with the substrate, other molecules and electron donors. The coupling to other magnetic species could provide another pathway of manipulating the spin structure of these molecules. A number of interesting studies on the interaction between magnetic impurities have already been published, using STM and the Kondo effect as a probe. The Kondo resonance of a Co atom on Cu_2N is split by the anti-ferromagnetic coupling with a nearby Fe atom, analogue to the effect of an external magnetic field [20]. For Co_2 dimers on Au(111), the disappearance of the Kondo resonance has been explained by the reduced exchange coupling between gold conduction electrons and ferromagnetic cobalt dimers [21]. For Ni/Au(111) T_K was observed to depend on the distance between two Ni atoms [22]. The broadening of the Kondo resonance was

132   6 Doping of MePc: Alkali and Fe Atoms

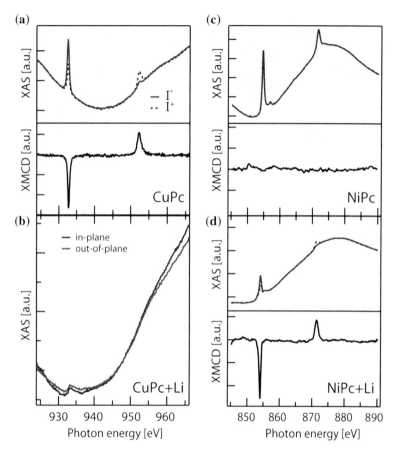

Fig. 6.15 X-ray absorption spectra (XAS) of undoped and doped CuPc (**a** and **b**) and NiPc (**c** and **d**) monolayers deposited on Ag(100), measured at the L edge of Cu and Ni. The x-ray magnetic dichroism (XMCD) curve is derived from subtraction of XAS spectra with opposite circular polarization. The addition of Li quenches the magnetic moment in CuPc and induces it on NiPc. Note that the spectra shown in (**b**) is the only one obtained with linearly polarized, but the absence of any 3d peak of the Cu ion already confirms the quenching of the magnetic moment in the doped CuPc layer. Measurements taken at T = 8 K, normal incidence and 5 T applied magnetic field at the ID08 beamline of the ESRF

used to determine the strength and sign of the coupling between two Co atoms on Cu(100) [23]. As another example, the coupling between Co atoms through atomic Cu chains was studied. Depending on the chain length T_K was found to be modulated by the RKKY interaction [24].

In CuPc the itinerant spin of the organic ligand could act as a coupling between the localized moment of the central ion and the external one of a magnetic adatom nearby. NiPc, with a similar delocalized moment, but without the magnetic ion, serves to disentangle the effects of the ligand and the ion. We codeposited Fe with

6.2 Doping with Fe Atoms

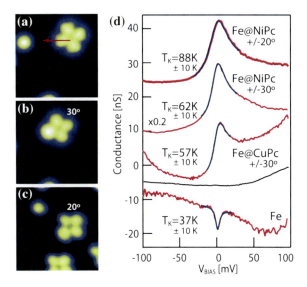

Fig. 6.16 **a** STM topography of a CuPc and an Fe atom (0.1 nA, −0.1 V, 5.1 nm × 5.1 nm). **b** Topography of the same region after a lateral manipulation of the CuPc towards the Fe, forming an Fe@CuPc complex. **c** Topography of Fe@NiPc with an azimuthal orientation of 20° with respect to the [011] surface axis (0.1 nA, −0.1 V, 5.1 nm × 5.1 nm). **d** dI/dV spectra taken on Fe (1.5 nA, −0.1 V) and on Fe within the Fe@NiPc (2.0 nA, −0.1 V) and Fe@CuPc (1.5 nA, −0.1 V) complexes shown in *red*. For Fe@CuPc a spectra representative for the other benzene rings is shown in *black*. The indicated T_K values are obtained from fitting the peak with a Fano function. The error is estimated by performing fits with different energy ranges for the same data set

CuPc and NiPc on a Ag(100) surface with in two separate preparations and studied the possible coupling between the magnetic moments of the molecules and the Fe atoms. We observe no evidence of direct magnetic coupling between the Fe and the ligand spin, nor the Cu ion. However, we do observe a change in their Kondo behavior. For Fe atom we see an increase of the Kondo temperature and a change in the Fano line shape, while for the molecules the Kondo resonance disappears entirely.

Fe atoms on the Ag(100) surface show a Kondo interaction with the conduction electrons of the metal at 5 K, an effect that has been observed for other magnetic adatoms as well [25–27]. Fitting a Fano lineshape to the characteristic zero bias peak, and subtracting the contribution of the thermal broadening due to the sample temperature of 5 K [28], we obtain a Kondo temperature of $T_K = 37 \pm 10$ K (see Fig. 6.16d). The value without the correction for the thermal broadening of $T_K = 77$ K is close to the also uncorrected $T_K = 80$ K found for Fe on Cu(100), measured at 5 K [26].

The in situ, low temperature deposition of a few Fe atoms on a Ag(100) surface containing single NiPc or CuPc molecules leads to several interacting configurations between Fe and the molecules. The most common configuration is shown in Fig. 6.16b. One Fe atom is directly interacting with one of the benzene rings of the molecule, seen by the brighter appearance in the topographic images. Lateral manip-

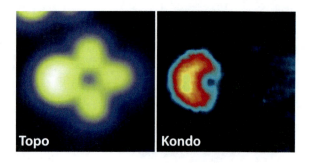

Fig. 6.17 STM topography of an Fe@NiPc type B complex and the corresponding dI/dV map of the Kondo resonance recorded at the same time (0.2 nA, 3 mV, 2.8 nm × 2.8 nm)

ulation allows us to verify this composition by pushing a MePc towards a single Fe atom, resulting in the creation of this type B (benzene) interaction (see Fig. 6.16a, b). The azimuthal orientation of the molecules can change when the Fe is interacting with them. Apart form the known ±30°, we observe a ±20° rotation from the [011] surface axis.

This type of interaction was observed both for NiPc and CuPc equally, with an identical effect on the spectroscopy. The Kondo resonance delocalized over the whole ligand and the inelastic vibrational excitations found for the unperturbed molecules disappear completely. The bright benzene ring with the Fe atom shows a Kondo resonance (see Fig. 6.16d). The corresponding Kondo temperature depends on the azimuthal angle of the molecule: For ±30°, it has increased to $T_K = 62 \pm 10$ K for Fe@NiPc, and to 57 ± 10 K for Fe@CuPc, while for ±20° Fe@NiPc we find $T_K = 88 \pm 10$ K.

The spatial distribution of this Kondo resonance is localized around the benzene ring containing the Fe, with an asymmetrical distribution that peaks on the part of the benzene farthest away from the center (see Fig. 6.17). This points towards the Fe atom as the origin of the observed resonance, while the ligand Kondo of the molecule is quenched. Moreover, the type of central ion has no effect, implying that the extra magnetic moment on the Cu^{2+} ion does not couple with the moment of the Fe. We therefore attribute the change in T_K solely to the interaction of the Fe adatom with the ligand.

In addition to the increase in T_K we observe a change in the Fano line shape, from a dip to a peak. This can be associated with different interference conditions between the tunneling into Ag conduction and Fe d states (see page 47). In this case the contribution of the Ag electrons decreases, a tunneling channel which could be inhibited by the presence of the molecule.

The orientation of the molecule has a noticeable effect on T_K, as summarized in Table 6.2. The change in the Kondo temperature of Fe could be related to a change in the local environment, caused by a different adsorption position or the interaction with molecular orbitals. Previous studies have shown that T_K for single adatoms/molecules is extremely sensitive to the local environment, leading to either an increase or a decrease of T_K [27, 29, 30]. If one plots the adsorption geometry of these complexes on Ag(100), assuming that the adsorption site of the TM-ion does

6.2 Doping with Fe Atoms

Table 6.2 Summary of the encountered species of Fe@NiPc/CuPc: The azimuthal orientation (φ) of the molecular axis with respect to the [011] direction of the substrate is given, as well as the Kondo temperatures obtained by a Fano function fit of the STS data

	φ	T_K [K]
Type B Fe@CuPc	±30°	57 ± 10
Type B Fe@NiPc	±30°	62 ± 10
Type B Fe@NiPc	±20°	88 ± 10
Fe close to NiPc	±30°	82 ± 10
Fe	–	37 ± 10

The error is estimated by performing fits with different energy ranges for the same data set

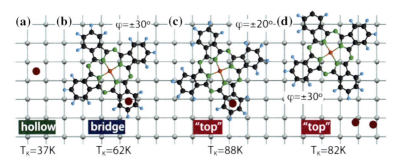

Fig. 6.18 Proposed geometrical model for the Fe@MePc complexes: The encountered Fe@MePc types and their respective azimuthal orientations are plotted. Assuming that the Fe is at the center of the benzene ring and that the adsorption site of the molecule is the same as the undoped, the difference in the Fe adsorption sites can be visualized. **a** A single Fe atom. **b, c** The type B complex with different azimuthal orientations. **d** The position of the Fe atoms close to NiPc that show a change in the Kondo behavior (see Fig. 6.21). The positions of the latter are extracted from topographic images by the distance and angle from the center of the NiPc

not change, a pattern emerges. Based on observations for Co on Cu(100) the single Fe atom can be assigned to a hollow adsorption site [25]. For the Fe atom within the complex, its position with respect to the substrate changes depending on the rotation of the molecule (see Fig. 6.18). Within this model we can identify three different Kondo temperatures for different adsorption sites of the Fe atom: $T_K = 37$ K for the hollow site, $T_K = 57$ K for the bridge site, and $T_K = 62$ K for an adsorption close to the top site.

Using lateral manipulation a second Fe atom can be inserted into a molecule. The azimuthal orientation changes during this process and a range of different configurations were observed. All of these 2Fe@MePc complexes show a Kondo effect only at the position of the Fe atoms. The line shape of the Kondo resonance is identical to those found in MePc with a single Fe. Likewise T_K lies in the same range from 37 K to 120 K. No clear dependency of T_K on the number of Fe atoms could be found. Moreover the two complexes shown in Fig. 6.19, with the Fe in neighboring or opposing benzene rings have the same Kondo temperature within the error of the measurement. Based on these facts we conclude that the two Fe inside the molecule

136 6 Doping of MePc: Alkali and Fe Atoms

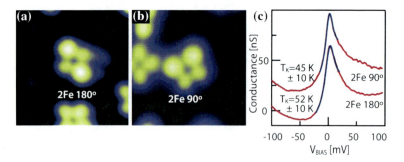

Fig. 6.19 **a, b** STM topography of two different 2Fe@NiPc configurations (0.1 nA, −0.1 V, 5.1 nm× 5.1 nm), **c** dI/dV spectra showing the corresponding Kondo resonances, including a Fano function fit (2 nA, −0.1 V)

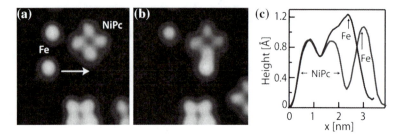

Fig. 6.20 **a, b** STM topography showing the lateral manipulation to move an Fe atom into the position where it interacts with a NiPc (0.1 nA, −0.1 V, 5.1 nm × 5.1 nm). **c** Height profiles of an Fe atom (*red*) in (**a**) and the Fe close to the NiPc (*blue*) in (**b**)

do not couple. The variations found in the Kondo temperature could be related to changes in the adsorption site of the Fe induced by the interaction with the molecule, in the same way observed for singly doped complexes. In the case of doubly doped complexes, the multiple azimuthal angles found within each type makes the correlation between T_K and a given adsorption site difficult to track.

We now turn to the other, less common type of interaction observed. An Fe atom that is close enough to a NiPc starts to interact with it. From topographic images, comparing the distance between the center of the Fe and the center of the NiPc, this critical distance can be estimated to be 1.1–1.3 nm, i.e. one or two lattice sites from the nearest H atoms of the molecule. The apparent height of the Fe changes and they appear brighter than the non-interacting ones (see Fig. 6.20). It is possible to create the type B interaction by pushing the Fe further towards the molecule (Fig. 6.21b).

The spectroscopy is tip position dependent as illustrated by the series of spectra taken along an Fe close to NiPc shown in Fig. 6.21a. Despite the presence of the Fe atom, the molecule exhibits the typical features found for unperturbed molecules, i.e. the Kondo resonance with T_K, and the vibrational excitation spectra. On the Fe atom

6.2 Doping with Fe Atoms

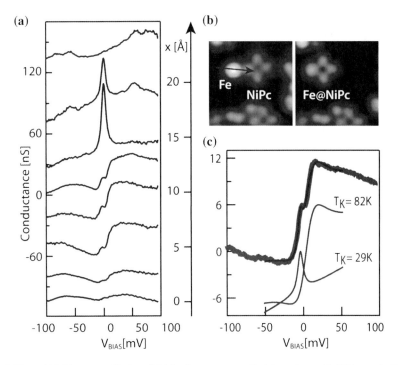

Fig. 6.21 **a** dI/dV spectra (2 nA, −0.1 V) taken across an Fe interacting with NiPc on Ag(100), along the red *arrow* shown in (**b**). The spectra on the *bottom* was taken on the *left side* of the *arrow*. **b** STM topography showing an interacting Fe close to NiPc (*left*). On the *right* the same molecule is shown after successfully pushing the Fe into the molecule to produce a type B complex (0.1 nA, −0.1 V, 4.8 nm × 4.8 nm). **c** dI/dV spectra taken on the Fe was fitted with two Fano functions. For one of them the $T_K = 29$ K was fixed, because it is the T_K value found on the other benzene rings

we observe a step like feature, combined with a peak at E_F. Using two overlapping Fano line shapes the feature can be fitted accurately (see Fig. 6.21c). For one of the Fano functions, we fix the line shape and $T_K = 29$ K of the unperturbed NiPc/Ag(100) based on the spectra taken on the other benzene rings. The second Fano function yields a step-like shape with Kondo temperature of $T_K = 82$ K. This value coincides with other T_K values found for the type B complexes. The two Fano functions could be explained by the overlapping of two non-coupled Kondo resonances. Spin-spin interaction most likely do not take place, based on the unchanged Kondo features found on the molecule.

The change in T_K for Fe could be caused by a change in the local environment, and is similar to that found in type B complexes corresponding to the "top" position. Indeed if we estimate the adsorption site of the Fe atom by measuring the distance and angle between its center and the MePc center for the two complexes we studied, they would be close to a "top" site (see Fig. 6.18d). The Fano line shape is an intermediate situation between the isolated Fe and that inside the molecule, which might result from a gradual reduction of the tunneling to conduction Ag states as the molecule gets closer.

6.3 Summary

Li-Doping

We have investigated the doping of MePcs (Me = Fe, Co, Ni, Cu) with charge donor (Li) adatoms. For the electron doping we find:

Site-specific charging in CuPc: Depending on the Li-CuPc bonding, charge is transferred to either the ligand or the TM. This mechanism allows to specifically control the spin and charge state of the CuPc, where we go from the (Q | $S_{total} = S_d + S_\pi$) = (1 | 1/2 + 1/2) state found for the undoped molecule on Ag, to a (2 | 1/2 + 0) in the type L complex, and to a (1 | 0 + 0) in the type M. This orbital selectivity is not observed for NiPc due to the lower affinity of the metal to accept charges. On the contrary, no charge transfer to the ligand states has been observed for FePc and CoPc.

TM dependence of the charge stability at the molecules: The type of TM ion determines, whether a Q = 1 or a Q = 2 configuration is stabilized. The Li@CuPc-L complex readily accepts a two electrons one from the Ag and the other from the Li. Li@NiPc-L, in contrast, is in a Q = 1 state. This stabilization of charge states seems to be in a delicate balance: Using a dopant with a lower ionization potential, such as K facilitates a Q = 2 configuration even for NiPc. On the other hand, a slightly changed adsorption geometry can reduce the charge transfer from 2 to 1 electron in Li@CuPc.

Atom-by-atom charge doping: Multiple doping of CuPc with up to 5 Li leads to a maximum transfer of 2 electrons, due to the splitting of the degeneracy of the $2e_g$ state and symmetry constraints imposed by the adsorption.

Short and long range Coulomb repulsion: The electronic structure of Li@CuPc-L and Li@NiPc-L complexes is affected by the presence of Li ions. A short range electrostatic interaction between the positively charged Li ions and Li@CuPc-L shifts the α_{2eg} orbital to lower energies. On the other hand long range interaction based on the Li induced charge redistribution in the substrate affects the effective intraorbital Coulomb repulsion in Li@NiPc, shifting the double occupation state (α_{2eg}+U) to higher energies.

Electrostatic stabilization of the type M configuration: With increasing Li coverage on a monolayer of CuPc, the most stable configuration is switched from the L to the M type, where the charge is transferred to the metal d states possibly due to electrostatic interaction between the Li-ions.

In general we have shown that electron doping is a promising pathway to tune the charge and spin of MePc, especially when they are susceptible to accept charge in both metal and ligand. We further find that on the surface not only charge transfer, but also electrostatic repulsion plays an important role.

6.3 Summary

Fe-Doping

We have further investigated the doping of MePcs spin carrier (Fe) adatoms:

Magnetic doping with Fe atoms: We have studied the interaction between Fe atoms and NiPc/CuPc molecules on Ag(100). No direct evidence of coupling between the magnetic moment of Fe and MePc was found. Instead, the MePc molecules change the local environment of the Fe atoms resulting in variations in the Kondo temperature and Fano line shape. A single molecule can embed up to 2 Fe atoms, which seem to remain uncoupled. The observed T_K was however affected by the multiple azimuthal orientations found for these complexes, probably also related to a change in the Fe adsorption site. A more indirect, apparently substrate mediated interaction has also been observed, where Fe adatoms at a distance of one or two lattice sites from the nearest H atom of the molecule exhibit a modified T_K and Fano line shape. We tentatively relate such variations to sa molecule induced change of the Fe adsorption site.

The changes in the Fe substrate coupling modulated by the presence of MePcs indicate that these kind of molecules could be used to tune the interaction of magnetic adatoms with metal surfaces.

References

1. P.J. Benning, J.L. Martins, J.H. Weaver, L.P.F. Chibante, R.E. Smalley, Electronic states of KxC60: insulating, metallic, and superconducting character. Science **252**(5011), 1417–1419 (1991). doi:10.1126/science.252.5011.1417
2. R.C. Haddon, A.F. Hebard, M.J. Rosseinsky, D.W. Murphy, S.J. Duclos, K.B. Lyons, B. Miller, J.M. Rosamilia, R.M. Fleming, A.R. Kortan, S.H. Glarum, A.V. Makhija, A.J. Muller, R.H. Eick, S.M. Zahurak, R. Tycko, G. Dabbagh, F.A. Thiel, Conducting films of C_{60} and c_{70} by alkali-metal doping. Nature **350**(6316), 320–322 (1991). doi:10.1038/350320a0
3. R. Yamachika, Controlled atomic doping of a single $C_6 0$ molecule. Science **304**(5668), 281–284 (2004). doi:10.1126/science.1095069
4. P. Gargiani, A. Calabrese, C. Mariani, M.G. Betti, Control of electron injection barrier by electron doping of metal phthalocyanines. J. Phys. Chem. **114**(28), 12258–12264 (2010). doi:10.1021/jp103946v
5. A. König, F. Roth, R. Kraus, M. Knupfer, Electronic properties of potassium doped FePc from electron energy-loss spectroscopy. J. Chem. Phys. **130**(21), 214503 (2009). doi:10.1063/1.3146812
6. V. Aristov, O. Molodtsova, M. Knupfer, Potassium doped Co phthalocyanine films: charge transfer to the metal center and the ligand ring. Organ. Electron. **12**(2), 372–375 (2011). doi:16/j.orgel.2010.12.003
7. M.F. Craciun, S. Rogge, A.F. Morpurgo, Correlation between molecular orbitals and doping dependence of the electrical conductivity in Electron-Doped Metal-Phthalocyanine compounds. J. Am. Chem. Soc. **127**(35), 12210–12211 (2005). doi:10.1021/ja054468j
8. M. Liao, S. Scheiner, Electronic structure and bonding in metal phthalocyanines, Metal=Fe Co., Ni, Cu, Zn, Mg. J. Chem. Phys. **114**(22), 9780 (2001). doi:10.1063/1.1367374
9. F. Roth, A. König, R. Kraus, M. Knupfer, Potassium induced phase transition of FePc thin films. J. Chem. Phys. **128**(19), 194711 (2008). doi:10.1063/1.2920179
10. K. Flatz, M. Grobosch, M. Knupfer, The electronic properties of potassium doped copper-phthalocyanine studied by electron energy-loss spectroscopy. J. Chem. Phys. **126**(21), 214702 (2007). doi:10.1063/1.2741539

11. L. Giovanelli, P. Vilmercati, C. Castellarin-Cudia, J. Themlin, L. Porte, A. Goldoni, Phase separation in potassium-doped ZnPc thin films. J. Chem. Phys. **126**(4), 044709 (2007). doi:10.1063/1.2432115
12. A. Calabrese, L. Floreano, A. Verdini, C. Mariani, and M. G. Betti, Filling empty states in a CuPc single layer on the Au(110) surface via electron injection. Phys. Rev. B **79**(11), 115446 (2009). doi:10.1103/PhysRevB.79.115446
13. H. Ding, K. Park, K. Green, Y. Gao, Electronic structure modification of copper phthalocyanine (CuPc) induced by intensive na doping. Chem. Phys. Lett. **454**(4–6), 229–232 (2008)
14. F. Evangelista, R. Gotter, N. Mahne, S. Nannarone, A. Ruocco, P. Rudolf, Electronic properties and Orbital-Filling mechanism in Rb-Intercalated copper phthalocyanine. J. Phys. Chem. **112**(16), 6509–6514 (2008). doi:10.1021/jp710197c
15. X.H. Qiu, G.V. Nazin, W. Ho, Vibronic states in single molecule electron transport. Phys. Rev. Lett. **92**(20), 206102 (2004). doi:10.1103/PhysRevLett.92.206102
16. S. Kera, H. Yamane, I. Sakuragi, K.K. Okudaira, N. Ueno, Very narrow photoemission bandwidth of the highest occupied state in a copper-phthalocyanine monolayer. Chem. Phys. Lett. **364**(1–2), 93–98 (2002). doi:10.1016/S0009-2614(02)01302-7
17. S. Kera, H. Yamane, N. Ueno, First-principles measurements of charge mobility in organic semiconductors: valence hole-vibration coupling in organic ultrathin films. Prog. Surf. Sci. **84**(5–6), 135–154 (2009). doi:10.1016/j.progsurf.2009.03.002
18. V. Simic-Milosevic, M. Heyde, N. Nilius, M. Nowicki, H. Rust, H. Freund, Substrate-mediated interaction and electron-induced diffusion of single lithium atoms on ag(001). Phys. Rev. B **75**(19), 195416 (2007). doi:10.1103/PhysRevB.75.195416
19. R.D. Muiño, D. Sánchez-Portal, V.M. Silkin, E.V. Chulkov, P.M. Echenique, Time-dependent electron phenomena at surfaces. Proc. Natl. Acad. Sci. USA **108**(3), 971–976 (2011). doi:10.1073/pnas.1008517107
20. A.F. Otte, M. Ternes, S. Loth, C.P. Lutz, C.F. Hirjibehedin, A.J. Heinrich, Spin excitations of a Kondo-Screened atom coupled to a second magnetic atom. Phys. Rev. Lett. **103**(10), 107203 (2009). doi:10.1103/PhysRevLett.103.107203
21. W. Chen, T. Jamneala, V. Madhavan, M.F. Crommie, Disappearance of the Kondo resonance for atomically fabricated cobalt dimers. Phys. Rev. B **60**(12), R8529–R8532 (1999). doi:10.1103/PhysRevB.60.R8529
22. V. Madhavan, T. Jamneala, K. Nagaoka, W. Chen, J. Li, S.G. Louie, M.F. Crommie, Observation of spectral evolution during the formation of a Ni_2 Kondo molecule. Phys. Rev. B **66**(21), 212411 (2002). doi:10.1103/PhysRevB.66.212411
23. P. Wahl, P. Simon, L. Diekhöner, V.S. Stepanyuk, P. Bruno, M.A. Schneider, K. Kern, Exchange interaction between single magnetic adatoms. Phys. Rev. Lett. **98**(5), 056601 (2007). doi:10.1103/PhysRevLett.98.056601
24. N. Néel, R. Berndt, J. Kröger, T.O. Wehling, A.I. Lichtenstein, M.I. Katsnelson, Two-site Kondo effect in atomic chains. Phys. Rev. Lett. **107**(10), 106804 (2011). doi:10.1103/PhysRevLett.107.106804
25. N. Knorr, M.A. Schneider, L. Diekhöner, P. Wahl, K. Kern, Kondo effect of single Co adatoms on Cu surfaces. Phys. Rev. Lett. **88**(9), 096804 (2002). doi:10.1103/PhysRevLett.88.096804
26. P. Wahl, Local Spectroscopy of correlated electron systems at metal surfaces. Ph.D. thesis, Universität Konstanz, 2005
27. M. Schneider, L. Vitali, P. Wahl, N. Knorr, L. Diekhöner, G. Wittich, M. Vogelgesang, K. Kern, Kondo state of Co impurities at noble metal surfaces. Appl. Phys. A **80**(5), 937–941 (2005). doi:10.1007/s00339-004-3119-7
28. K. Nagaoka, T. Jamneala, M. Grobis, M.F. Crommie, Temperature dependence of a single Kondo impurity. Phys. Rev. Lett. **88**(7), 077205 (2002). doi:10.1103/PhysRevLett.88.077205
29. L. Gao, W. Ji, Y. B. Hu, Z. H. Cheng, Z. T. Deng, Q. Liu, N. Jiang, X. Lin, W. Guo, S. X. Du, W. A. Hofer, X. C. Xie, and H. Gao, Site-Specific Kondo effect at ambient temperatures in Iron-Based molecules. Phys. Rev. Lett. **99**(10) (2007). doi:10.1103/PhysRevLett.99.106402
30. N. Néel, J. Kröger, R. Berndt, T.O. Wehling, A.I. Lichtenstein, M.I. Katsnelson, Controlling the Kondo effect in $CoCu_n$ clusters atom by atom. Phys. Rev. Lett. **101**(26), 266803 (2008). doi:10.1103/PhysRevLett.101.266803

Chapter 7
Conclusions and Outlook

The application of molecules in technological devices depends crucially on the understanding of their behavior on metallic electrodes or substrates. Gas phase molecular electronic and magnetic properties are often times modified or quenched upon adsorption. The results presented in this thesis provide a comprehensive overview of the influence of molecule-substrate and molecule-molecule interactions on the electronic and magnetic structure of molecules adsorbed on a metallic substrate. The main mechanisms involved are hybridization of molecular and substrate states, and charge transfer. In addition we have explored how to modify charge (Q) and spin (S) by doping them with single atoms. Our model system were MePc, metal-organic complexes consisting of an organic ligand and a central metal ion, adsorbed on a Ag(100) surface. Investigating four different kinds of MePcs (Me =Fe, Co, Ni, Cu) allowed us to disentangle the role of the central ion and the ligand.

The interaction between the MePc and the surface is closely related to the adsorption geometry of individual molecules. In Chap. 4 we found that the adsorption of MePc molecules is driven by bond optimization between the ligand and the underlying Ag substrate. Due to this interaction, the flat lying single MePc molecules exhibit two possible equivalent orientations of the molecular axis on the surface. The resulting symmetry mismatch between molecule and substrate leads to an asymmetric hybridization of different molecular orbitals. As a result the achiral molecule show chiral contrast in STM topography, which is related to an electronic rather than a conformational perturbation of the molecule.

As the coverage is increased, intramolecular vdW forces become important, leading to a transfer of the electronic chirality of single molecules to the organizational level in clusters, independent of the central TM ion. Ostwald ripening of the originally racemic mixture of clusters, together with the thermally induced switching of chirality by azimuthal rotation, leads to spontaneous symmetry breaking, resulting in mesoscopic homochiral molecular layers. These enantiopure molecular layers and the orbital-selective chirality offer an interesting way to control the electronic properties and optical response of metal surfaces.

In Chap. 5 we discussed how the electronic and magnetic properties depend on the whole molecule (organic ligand and central metal ion). While all molecules receive approximately one electron from the surface, depending on the type of central ion, and the character of its frontier d orbitals, the consequences are different. For d states lying within the molecular plane, as is the case for NiPc/CuPc, the main hybridization channel is an organic π ligand orbital, leading to the acceptance of an extra electron here. This unpaired electron causes an additional magnetic moment, whereas the magnetic state of the central ions remains unperturbed. For CuPc, intramolecular exchange coupling between d and π electrons leads to the formation of a triplet (S = 1) ground state.

In contrast to this in FePc and CoPc the interaction through out-of-plane d states creates a complex charge reorganization and induces a mixed valence configuration. For both of these cases this interaction leads to a reduction of the ion's magnetic moment.

The influence of intermolecular interaction was the second topic in Chap. 5. For CuPc molecules we found that the partially occupied ligand orbital is strongly affected by neighboring molecules, evidenced by a complex evolution of the electronic structure. This is in contrast with the case of CoPc, where the substrate induced charge is localized more at the center of the molecule and hence appears to be unaffected by intermolecular interactions. The difference between the two cases could also originate from the stronger interaction between CoPc and the surface due to direct TM-Ag interaction.

In multilayers of CuPc we find a gradual vertical decoupling from the substrate. The molecules on the 2nd layer maintain the flat adsorption geometry, however experience a 45° azimuthal rotation. Their electronic structure appears to be in an intermediate state, sufficiently decoupled to show the gas-phase HOMO-LUMO gap separation and vibronic coupling, but still with some weak features within the gap. Starting form the 3rd layer, the molecules align their ligand axis and start to tilt. The electronic decoupling becomes complete, evidenced by a zero conductance gap, and second tunneling barrier.

In Chap. 6 we explored the possibilities of manipulating the electronic and magnetic characteristics of MePc by adding Li adatoms as electron donors. Through atom by atom doping of CuPc, we were able to selectively dope ligand and metal states by changing the alkali-molecule bonding configuration. Through this selectivity we were able to achieve a variety of Q and S configurations, reaching Q = 1 and 2 electrons, and S = 0, 1/2, and 1 with a single alkali dopant. This type of control is a step ahead for the future application of molecules in spintronic devices.

Lastly we have studied the addition of magnetic dopants, in the form of Fe atoms to NiPc and CuPc. We did not observe any direct evidence of a coupling between the magnetic moments of Fe, Cu, or the ligand. Instead the interaction with molecules changed the local environment of the Fe atom, and with it its Kondo temperature.

These findings are an important step in understanding the electronic and magnetic structure of metal-organic ligand complexes deposited on a surface. Clearly the molecule and its enviroment has to be considered as a whole system: The ion, the ligand, neighboring molecules and the surface have to be considered to understand

7 Conclusions and Outlook

the full molecular magnetic structure. The detailed step by step investigation (single molecule, cluster, monolayer, multilayer) allowed us to shed light on the evolution and the mechanisms behind the changes in the electronic and magnetic structure. The need for experiments in this area is also underlined, as effects such as the addition of ligand spins are difficult to predict from ab-initio calculations.

At several points during the course of this thesis questions were encountered that would make interesting follow up experiments. (i) The electronic structure found for the monolayer of CuPc insinuated a possible dispersion of the unoccupied states, probably due to hybridization with an upshifted surface state. A similar dispersion was observed for PTCDA molecules [1, 2]. The same state showed a strong Stark shift like behavior with the tip sample distance. An investigation of the evolution of this state as a function of island size and shape to find the dispersion might shed light into this problem. (ii) The quenching of the ligand spin Kondo interaction for CuPc was caused by the interplay between molecule-molecule and molecule-substrate interactions. Adsorption on different metallic surfaces or changes in the molecular end-groups, could result in a system where the both the charge transfer to the ligand and the Kondo screening persist. (iii) The observed change in the Kondo coupling between Fe atoms and Ag substrate mediated by the MePc suggested that these molecules could be used as templates to shape adatom-substrate interaction [3, 4].

Several pathways of continuing the systematical line of research started in this work can be proposed. The role of the substrate can explored in more detail. In this thesis a relative inert noble metal surface Ag(100) was used, but already the interaction with it lead to drastic changes in the molecules. Choosing substrates with either stronger or weaker interaction can help understanding further the role of the underlying substrate. Preliminary studies on Au(111) were done in this thesis (Appendix A).

Magnetic surfaces could add an additional interaction dimension. Metal-organic molecules have already been deposited on ferromagnetic samples, showing ferro and antiferromagnetic molecule-substrate coupling [5–9]. The ideal complement would be antiferromagnetic surfaces, which could lead to a stabilization of the magnetic moments through exchange bias effects [10, 11]. Possibilities include metallic films or oxide based anti-ferromagnetic films, which might help to disentangle the magnetic and charge transfer interactions.

Another pathway worth exploring would be extending the variety of the chemistry of the organic ligand part. The oxidation state of the central ion can be tuned by chemically doping with electron acceptor species, for instance $Cl_2Co(III)Pc$ [12]. Further by functionalizing the ligand part of MePc, the coupling to the substrate could be tuned, e.g., sulfonated or thiol-derivatized phthalocyanines would be strongly bonded to the Ag surface through the strong S-Ag affinity [13]. At the same time intermolecular interactions can be engineered. Polar end groups would lead to stronger hydrogen bond between molecules. By carefully choosing these groups a wide range of network geometries and interaction strengths are possible.

References

1. R. Temirov, S. Soubatch, A. Luican, F. S. Tautz, Free-electron-like dispersion in an organic monolayer film on a metal substrate. Nature **444**(7117), 350–353 (2006). doi:10.1038/nature05270
2. M. S. Dyer, M. Persson, The nature of the observed free-electron-like state in a PTCDA monolayer on Ag(111). New J. Phys. **12**(6), 063014 (2010). doi:10.1088/1367-2630/12/6/063014
3. F. Rosei, M. Schunack, P. Jiang, A. Gourdon, E. Lægsgaard, I. Stensgaard, C. Joachim, F. Besenbacher, Organic molecules acting as templates on metal surfaces. Science **296**(5566), 328–331 (2002). doi:10.1126/science.1069157
4. S. Li, H. Yan, L. Wan, H. Yang, B. H. Northrop, P. J. Stang, Control of supramolecular rectangle Self-Assembly with a molecular template. J. Am. Chem. Soc. **129**(30), 9268–9269 (2007). doi:10.1021/ja0733282
5. J. Brede, N. Atodiresei, S. Kuck, P. Lazicacute, V. Caciuc, Y. Morikawa, G. Hoffmann, S. Blügel, R. Wiesendanger, Spin- and energy-dependent tunneling through a single molecule with intramolecular spatial resolution. Phys. Rev. Lett. **105**(4), 047204 (2010). doi:10.1103/PhysRevLett.105.047204
6. M. Bernien, J. Miguel, C. Weis, M. Ali, J. Kurde, B. Krumme, P. Panchmatia, B. Sanyal, M. Piantek, P. Srivastava, K. Baberschke, P. Oppeneer, O. Eriksson, W. Kuch, H. Wende, Tailoring the nature of magnetic coupling of Fe-Porphyrin molecules to ferromagnetic substrates. Phys. Rev. Lett. **102**(4) (2009). doi:10.1103/PhysRevLett.102.047202
7. H. Wende, M. Bernien, J. Luo, C. Sorg, N. Ponpandian, J. Kurde, J. Miguel, M. Piantek, X. Xu, P. Eckhold, W. Kuch, K. Baberschke, P. M. Panchmatia, B. Sanyal, P. M. Oppeneer, O. Eriksson, Substrate-induced magnetic ordering and switching of iron porphyrin molecules. Nat. Mater. **6**(7), 516–520 (2007). doi:10.1038/nmat1932
8. A. Lodi Rizzini, C. Krull, T. Balashov, J. J. Kavich, A. Mugarza, P. S. Miedema, P. K. Thakur, V. Sessi, S. Klyatskaya, M. Ruben, S. Stepanow, P. Gambardella, Coupling single molecule magnets to ferromagnetic substrates. Phys. Rev. Lett. **107**(17), 177205 (2011). doi:10.1103/PhysRevLett.107.177205
9. N. Atodiresei, J. Brede, P. Lazicacute, V. Caciuc, G. Hoffmann, R. Wiesendanger, S. Blügel, Design of the local spin polarization at the Organic-Ferromagnetic interface. Phys. Rev. Lett. **105**(6), 066601 (2010). doi:10.1103/PhysRevLett.105.066601
10. J. Nogués, J. Sort, V. Langlais, V. Skumryev, S. Suriñach, J. Muñoz, M. Baró, Exchange bias in nanostructures. Phys. Rep. **422**(3), 65–117 (2005). doi:10.1016/j.physrep.2005.08.004
11. V. Skumryev, S. Stoyanov, Y. Zhang, G. Hadjipanayis, D. Givord, J. Nogues, Beating the superparamagnetic limit with exchange bias. Nature **423**(6942), 850–853 (2003). doi:10.1038/nature01687
12. J. F. Myers, G. W. R. Canham, A. B. P. Lever, Higher oxidation level phthalocyanine complexes of chromium, iron, cobalt and zinc. phthalocyanine radical species. Inorg. Chem. **14**(3), 461–468 (1975). doi:10.1021/ic50145a002
13. F. Dumoulin, M. Durmuş, V. Ahsen, T. Nyokong, Synthetic pathways to water-soluble phthalocyanines and close analogs. Coord. Chem. Rev. **254**(23–24), 2792–2847 (2010). doi:10.1016/j.ccr.2010.05.002

Appendix
CuPc on Au(111)

To study the influence of the substrate on the charge transfers to the $2e_g$, we also investigated the adsorption of MePc molecules on different metal surfaces. The results show that the type of surface plays a critical role on the molecular charge and spin state.

To investigate the influence of the surface on the charge transfer, we studied individual CuPc molecules on a Au(111) surface. Compared to Ag(100), this surface is less reactive because of the stronger noble-metal character of Au as well as the tighter packed structure of the (111) plane. This surface presents a more complex reconstruction and several different adsorption site exists [1]. In Fig.1a, a topographic image of the Au(111) surface showing the characteristic herring-bone reconstruction is shown. Molecules on both adsorption sites (fcc, hcp) present similar dI/dV spectra. Figure.1b shows the results for fcc, where the a_{1u} and $2e_g$ states can readily be identified. The dI/dV maps taken at these energies (-0.7 and 1.4 V) are in agreement with maps of $a_{1u}/2e_g$ orbitals measured for CoPc on the same surface [2]. The $2e_g$ orbital thus remains unfilled as in the gas-phase, consistent with the fact that no Kondo resonance is observed for this system.

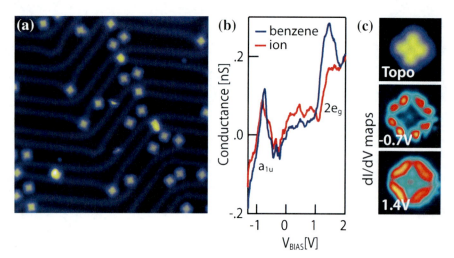

Fig .1 a STM topography of CuPc molecules on a Au(111) surface (0.1 nA, −1.2 V, 39 × 39 nm) **b** dI/dV spectra taken on CuPc on the fcc, *blue* on the benzene and *red* on the Cu ion. A spectrum acquired on the bare Au surface has been subtracted to enhance the molecular features (0.25 nA, −1.3 V) **c** dI/dV maps of the same molecule, showing the a_{1u} state at −0.7 V and the $2e_g$ at 1.4 V (3 × 3 nm)

References

1. L. Gao, W. Ji, Y.B. Hu, Z.H. Cheng, Z.T. Deng, Q. Liu, N. Jiang, X. Lin, W. Guo, S.X. Du, W.A. Hofer, X.C. Xie, H. Gao, Site-specific Kondo effect at ambient temperatures in iron-based molecules. Phys. Rev. Lett. **99**(10), 106402 (2007). doi:10.1103/PhysRevLett.99.106402
2. Z. Li, B. Li, J. Yang, J.G. Hou, Single-molecule chemistry of metal phthalocyanine on noble metal surfaces. Acc. Chem. Res. **43**(7), 954–962 (2010). doi:10.1021/ar9001558

Printed by Publishers' Graphics LLC
DBT131117.20.08.150